Astrophysik seit 1900
Jubiläum von Karl Schwarzschild (1873–1916)
und Ejnar Hertzsprung (1873–1967)

Astrophysics since 1900
Jubilee of Karl Schwarzschild (1873–1916)
and Ejnar Hertzsprung (1873–1967)

Abbildung 0.1:
Karl Schwarzschild & Ejnar Hertzsprung in Göttingen (1909)
Cod. Ms. K. Schwarzschild 23: 1,13 (Nachlass K. Schwarzschild; SUB Universität Göttingen)

Nuncius Hamburgensis
Beiträge zur Geschichte der Naturwissenschaften
Band 59

Wolfschmidt, Gudrun (Hg.)

Astrophysik seit 1900

Jubiläum von Karl Schwarzschild (1873–1916) und Ejnar Hertzsprung (1873–1967)

Proceedings der Tagung des
Arbeitskreises Astronomiegeschichte
in der Astronomischen Gesellschaft in Berlin 2023.

Astrophysics since 1900 – Jubilee of Karl Schwarzschild (1873–1916) and Ejnar Hertzsprung (1873–1967)

Ahrensburg bei Hamburg: tredition 2024

Nuncius Hamburgensis
Beiträge zur Geschichte der Naturwissenschaften

Hg. von Gudrun Wolfschmidt, Universität Hamburg,
AG Geschichte der Naturwissenschaft und Technik
(ISSN 1610-6164).

Dieser Titel wurde inspiriert von „Sidereus Nuncius“ und von „Wandsbeker Bote“.

Wolfschmidt, Gudrun (Hg.): Astrophysik seit 1900 – Jubiläum von Karl Schwarzschild (1873–1916) und Ejnar Hertzsprung (1873–1967). Astrophysics since 1900 – Jubilee of Karl Schwarzschild (1873–1916) and Ejnar Hertzsprung (1873–1967). Proceedings der Tagung des Arbeitskreises Astronomiegeschichte in der Astronomischen Gesellschaft in Berlin, Sept. 2023. Ahrensburg: tredition (Nuncius Hamburgensis – Beiträge zur Geschichte der Naturwissenschaften, Band 59) 2024.

Cover vorne und Frontispiz: Karl Schwarzschild & Ejnar Hertzsprung in Göttingen (1909), (Cod. Ms. K. Schwarzschild 23: 1,13, Nachlass K. Schwarzschild; SUB Universität Göttingen)
Cover hinten: HRD (Riesen- und Zwergsterne) (CC3, HR-sparse-de.svg, Rursus)

AG Geschichte der Naturwissenschaft und Technik, Hamburger Sternwarte,
Bundesstraße 55 – Geomatikum, 20146 Hamburg, Germany
`https://www.fhsev.de/Wolfschmidt/GNT/home-wf.htm`,
`gudrun.wolfschmidt@uni-hamburg.de`

Dieser Band wurde gefördert von der Hans Schimank-Gedächtnisstiftung und dem *Arbeitskreis Astronomiegeschichte in der Astronomischen Gesellschaft.*

Verlag: tredition GmbH, An der Strusbek 10, 22926 Ahrensburg, Germany
ISBN – 978-3-384-44634-3 (Softcover), 978-3-384-44635-0 (Hardcover),
978-3-384-44636-7 (e-Book), © 2024 Gudrun Wolfschmidt.

Inhaltsverzeichnis

Abbildung 0.2:
Brandenburger Tor Berlin
(Foto: Gudrun Wolfschmidt)

Vorwort – Preface

Astrophysik seit 1900 – Astrophysics since 1900

Wolfschmidt, Gudrun (Hamburg)

Astrophysik seit 1900 – Astrophysics since 1900 – this conference of the Arbeitskreis Astronomiegeschichte in der Astronomischen Gesellschaft took place in Bremen from 10–11 September 2023.[1]

Introduction: From classical astronomy to theoretical astrophysics

The topic, astrophysics since 1900, is outlined by way of introduction, and the development from the end of the 19th century to the middle of the 20th century is presented in eight chapters.

Until the 19th century, classical astronomical research was mainly based on measuring stellar and planetary positions for compiling star catalogues, and astronomers were mathematicians.

Around 1860, astronomy underwent a revolution. Instead of only studying only the direction of star light, the quantity and quality of radiation were studied for the first time. This was the beginning of modern (observational) *"astrophysics"*; this term was introduced by Johann Karl Friedrich Zöllner (1834–1882). The astrophysicists began to investigate the properties of the celestial bodies with physical and chemical methods. The new topics were photometry, astrophotography, spectroscopy / spectralanalysis, and solar physics.

Astrophysics in the 1st half of the 20th century

The first article focuses on the 150th anniversary of Schwarzschild and Hertzsprung. Karl Schwarzschild (1873–1916) was the pioneer of *theoretical astrophysics* around 1900. He introduced new theoretical topics like solar physics,

1 Website of the conference of the AKAG: `https://www.fhsev.de/Wolfschmidt/events/akag-berlin-2023.php`. There you will also find the link to the Booklet of Abstracts: `https://www.fhsev.de/Wolfschmidt/events/pdf/Booklet-AKAG-Berlin-2023-Abstract+Cover.pdf`.

theory of stellar atmospheres, stellar structure, and Einstein's *General Theory of Relativity.* He achieved the breakthrough of the *Hertzsprung-Russell-Diagram* (HRD), developed independently by Ejnar Hertzsprung (1873–1967) in 1905, and Henry Norris Russell (1877–1957) in 1910/12. The HRD is essential for the discussion of stellar evolution.

Schwarzschild was convinced that the interplay with other areas of exact sciences was of great importance for the development of astrophysics.

Ejnar Hertzsprung (1873–1967), after having studied studied chemical engineering, was interested in photochemistry. In 1905 and 1907, Hertzsprung published now classic articles *Zur Strahlung der Sterne* (On the Radiation of Stars) about his attempts to measure the luminosity of stars on the basis of their spectra. 50 years after Kirchhoff and Bunsen, the development of astrophysics reached a peak with Schwarzschild and Hertzsprung. Modern theoretical astrophysics began through the inclusion of important physical topics in astronomy.

The contribution of Markus Bautsch *Die von Johann Jakob Balmer (1825–1898) gefundenen Zahlenverhältnisse bei den Spektrallinien des Wasserstoffs* is dedicated to the reception of these new ideas (1884), which astonished and perplexed the world likewise. It was not until 1926 that the mystery of the Balmer series was solved with the help of quantum mechanics.
Very interesting is the information that through the friendship of the mathematician and composer Hans Sommer (1837–1922) in the early 1890s with the young composer Richard Strauss (1864–1949) created a composition that used the frequencies of the Balmer series in 1893 to create the pitches of a composition (Till-Eulenspiegel motif).

Xian Wu presents Heinrich Kayser's (1853–1940) *"Handbuch der Spectroscopie" (1900–1934) und dessen Bedeutung für die Astrophysik* ("Handbook of Spectroscopy" (1900–1934) and its significance for astrophysics). This eight-volume work was enthusiastically received and favourably reviewed worldwide – by chemists, physicists and especially astrophysicists. This is characterised in the article by impressive quotes.

The article *Karl Schwarzschild and Ejnar Hertzsprung in Potsdam (1910–1916)* by Adriaan Raap is also dedicated to the topic of the 2023 anniversary. Schwarzschild 's trip to the USA in 1910 on the occasion of the *International Solar Union* meeting in Los Angeles is interesting. He not only visited all the major observatories, but in particular met Henry Norris Russell (1877–1957) at *Princeton University*, to whom he reported on Hertzsprung's research, which pointed to a relationship between the luminosity and colour temperature of a star, which eventually led to today's Hertzsprung-Russell diagram. Another, somewhat less well-known topic is meteorology and navigation of balloons and zeppelins, which led to a weather station in the Belgian Ardennes in 1914, where Schwarzschild was active in the First World War. However, he also remained constantly connected to current physics, and made remarkable contributions to Einstein's general theory of relativity.

The article by Maik Schmerbauch sheds light on *Hans Kienle (1895–1975) – Scientist between astrophysics and politics in the 20th century*, based on archive material. The stages of his scientific life in various observatories took place under different political systems.

The next article by Stefan L. Wolff deals with *displacement and emigration of astrophysicists under National Socialism* – a total of 16 dismissed scientists, a topic that has hardly been investigated so far.

Modern Astrophysics

Ralph N. & Dagmar L. Neuhäuser investigate *The rapid evolution of Betelgeuse through the Hertzsprung gap from a yellow to a red supergiant in historical times* (Die rasche Evolution von Beteigeuze durch die Hertzsprung-Lücke vom gelben zum roten Überriesen in historischer Zeit). Historical reports on the colours of stars are also compiled and critically evaluated. Due to the colour change of Betelgeuse to a red supergiant in the last millennia, the mass of Betelgeuse can be precisely determined by the historically documented colour change – and thus also its remaining lifetime until the supernova in only about 1.5 million years.

Finally, Dietrich Lemke reports on *Experienced history – The James Webb Space Telescope – From idea to mission* (Erlebte Geschichte – Das James-Webb-Space-Telescope – Von der Idee zur Mission).

Excursion

It was a successful conference, complemented by a visit to the Archenhold Observatory in Berlin-Treptow, founded in 1896 by Friedrich Simon Archenhold (1861–1939), cf. p. 205. The impressive large 68 cm-refractor (21 m focal length), the giant telescope, made by Steinheil of Munich, and by C. Hoppe of Berlin, is the *longest* fully movable refracting telescope in the world.

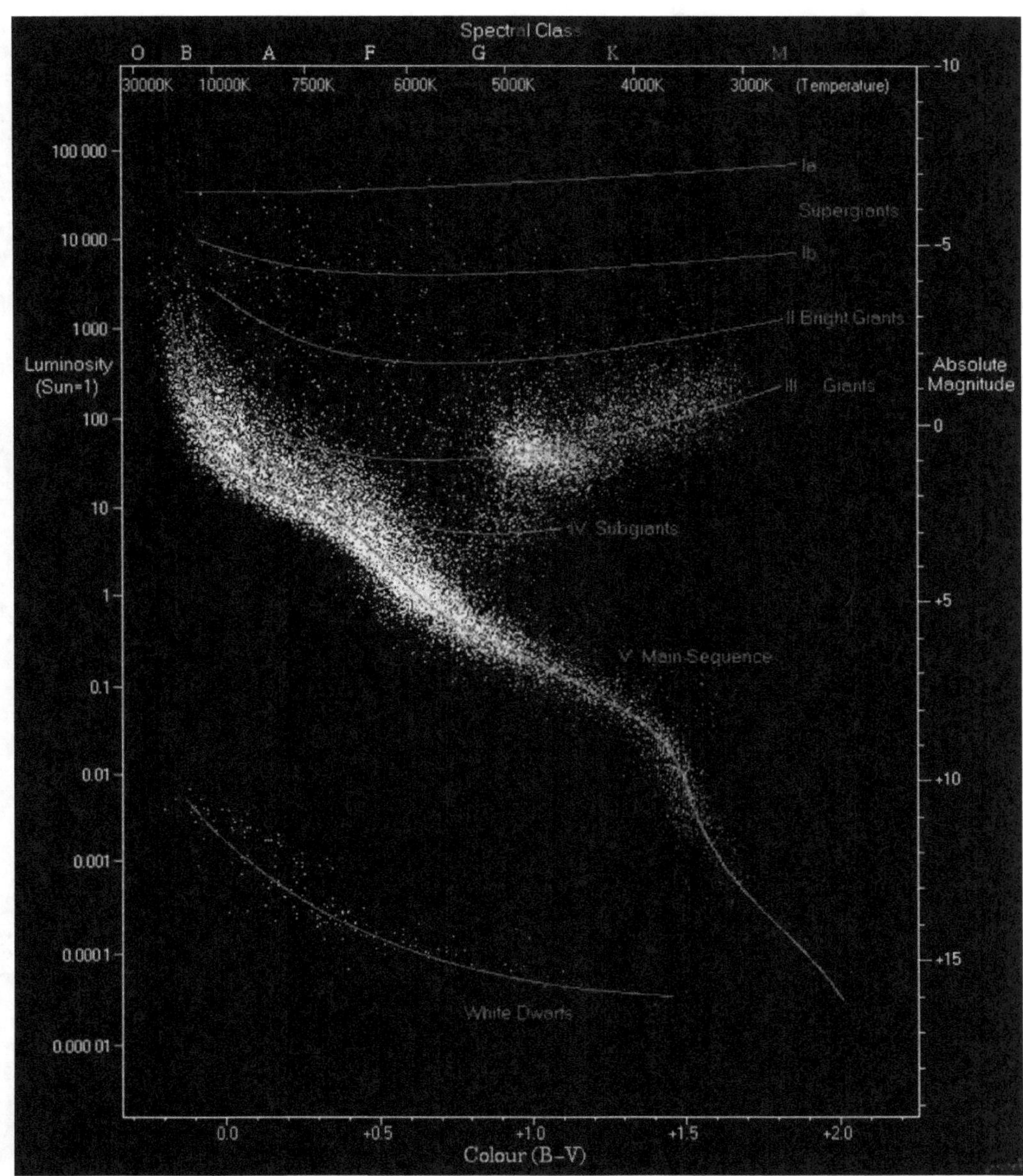

Figure 0.3:
An observational Hertzsprung-Russell diagramme with 22,000 stars plotted from the Hipparcos Catalogue and 1,000 from the Gliese Catalogue of nearby stars. One recognizes clearly the main sequence, the white dwarfs, and the giants.

(CC2.5, Richard Powell)

Abbildung 1.1:
Karl Schwarzschild (1873–1916), pioneer of astrophysics, and Black Hole with accretion disk

(© Leibniz-Institut für Astrophysik Potsdam (AIP))

Einführung: Astrophysik seit 1900 – zum Jubiläum von Schwarzschild und Hertzsprung

Gudrun Wolfschmidt (Hamburg)

Abstract: Astrophysics since 1900 – Jubilee of Schwarzschild and Hertzsprung

Karl Schwarzschild (1873–1916) began his astrophysical studies in Strasbourg in the field of variable stars. At the Kuffner Observatory in Vienna, he created photographic photometry. In Munich he devoted himself to the orbit determination of spectroscopic binaries and stellar statistics. As director of the Göttingen Observatory in 1901, he started theoretical optics and instrument development. Inspired by the total solar eclipse in Algeria (1905), he began working on solar physics, which led him to the theory of stellar atmospheres in 1906. As director of the Potsdam Astrophysical Observatory (1909 to 1916), Schwarzschild was particularly concerned with the emerging *General Theory of Relativity.*

Ejnar Hertzsprung (1873–1967), former professor in Göttingen from 1909, recognised that "giant stars" and "dwarf stars" can occur in stars with the same surface temperature, and as early as 1905 he conceived a colour-magnitude diagramme – decisive for stellar evolution. Henry Norris Russell (1877–1957) learned of this through Schwarzschild in 1910, and presented his diagram in 1912. Fifty years after Kirchhoff & Bunsen (1859), the development of astrophysics reached a climax with Schwarzschild and Hertzsprung; modern theoretical astrophysics began by incorporating important physical subfields into astronomy.

Zusammenfassung

Karl Schwarzschild (1873–1916) begann seine astrophysikalischen Studien in Straßburg auf dem Gebiet der Veränderlichen Sterne. In der Wiener Kuffner-Sternwarte schuf er die photographische Photometrie. In München widmete er sich der Bahnbestimmung spektroskopischer Doppelsterne und Stellarstatistik. Als Direktor der Sternwarte Göttingen 1901 widmete er sich der theoretischen Optik und Instrumenten-Entwicklung. Angeregt durch die Sonnenfinsternis in Algerien (1905) begann er mit Sonnenphysik, was ihn ab 1906 zur Theorie der Sternatmosphären führte. Als Direktor des Astrophysikalischen Observatoriums Potsdam (1909 bis 1916) beschäftigte sich Schwarzschild besonders mit der gerade entstehenden *Allgemeinen Relativitätstheorie.*

Ejnar Hertzsprung (1873–1967), ab 1909 a.o. Professor in Göttingen, erkannte er, dass bei Sternen gleicher Oberflächentemperatur „Riesensterne“ und „Zwergsterne“ auftreten können und konzipierte bereits 1905 ein Farben-Helligkeits-Diagramm – entscheidend für die Sternentwicklung. Henry Norris Russell (1877–1957) erfuhr durch Schwarzschild 1910 davon und stellte sein Diagramm 1912 vor. 50 Jahre nach Kirchhoff und Bunsen (1859) erreichte die Astrophysik seit 1900 mit Schwarzschild unter Mitwirkung von Hertzsprung einen Höhepunkt; durch Einbeziehung wichtiger physikalischer Teilgebiete in die Astronomie begann die moderne theoretische Astrophysik.

1.1 Introduction: The Rise of Astrophysics

Until the 19th century, classical astronomical research was based on measuring stellar and planetary positions for compiling star catalogues, in addition, since Newton's gravitational theory also celestial mechanics, and astronomers were mathematicians. Around 1860, astronomy underwent a revolution. Instead of only studying the direction of star light, the quantity and quality of radiation were studied for the first time. This was the beginning of modern (observational) *"astrophysics"*. The astrophysicists began to investigate the properties of the celestial bodies with physical and chemical methods. The new topics were photometry, photography, spectroscopy/spectralanalysis, and solar physics.

The next step was the introduction of *theoretical astrophysics* around 1900. Karl Schwarzschild (1873–1916), born 150 years ago (jubilee 2023), started with observational astrophysics by introducing photographic photometry. With his pioneering research as director of Göttingen and Potsdam observatories, he introduced new theoretical topics like solar physics, theory of stellar atmospheres, stellar structure, and Einstein's *General Theory of Relativity.* He achieved the breakthrough of the *Hertzsprung-Russell-Diagram* (HRD), developed independently by Ejnar Hertzsprung (1873–1967) in 1905, and Henry Norris Russell (1877–1957) in 1910/12. The HRD is essential for the discussion of stellar evolution. Modern theoretical astrophysics began through the inclusion of the important new physics (quantum theory) in astronomy.

1.1.1 Transition from Classical Astronomy to Observational Astrophysics

This topic was introduced and discussed by me since 2008, when I organized the ICOMOS conference *Cultural Heritage of Astronomical Observatories – From Classical Astronomy to Modern Astrophysics* in Hamburg. Here I would like to give only a short summary.

The emphasis in classical astronomy was on astrometry, timekeeping, especially for navigation, and surveying, and providing the time to the public clocks as well as to the timeballs, celestial mechanics, calculation of ephemeris and calendars. The most important topic of research until the first half of the 19th century was positional astronomy with meridian circles for compiling star catalogues by the observation of star transits with meridian circles.

Around 1860 astronomy underwent a revolution. In the context of "classical astronomy", only the direction of star light was studied. In the 1860s quantity and quality of radiation were studied for the first time. This was the beginning of modern "astrophysics". Simon Newcomb (1835–1909) wrote: *"that the age of great discoveries in any branch of science had passed by, yet so far as astronomy is concerned, it must be confessed that we do appear to be fast reaching the limits of our knowledge."* (Newcomb 1888). But he was wrong. In the second half of the 19th century a new, revolutionary branch of astronomy began to be practised – the "New Astronomy" – as Newcomb later called it, in contrast to positional astronomy.

The main topic of research had crossed over from classical astronomy to the new astrophysics. Concerning the change of the research field, Johann Karl Friedrich Zöllner (1834–1882) introduced the term "astrophysics" in 1865. Instead of combining astronomy with mathematics with compiling large star catalogues and the calculation of orbits of the planets and comets, astronomers – or better astrophysicists – began around 1860 to investigate the properties of celestial bodies with physical and chemical methods.

The new topics were photometry (1860), astrophotography (1839, 1880s), spectroscopy/spectral analysis (Fraunhofer 1814, Kirchhoff & Bunsen 1859), and solar physics. In the laboratory, the collected data were analysed and the photographic plates are measured. Julius Scheiner (1858–1913), one of the first professors at Astrophysical Observatory Potsdam, wrote about the research topics of astrophysics in 1890:[1]

> *"There can be no doubt that spectral analysis occupies the first place here and will continue to do so for the foreseeable future, all the more, so as the latter research has opened up a new field of observational activity that promises to yield extensive results in the future."*[2]

Astrophysics started with amateur astronomers in private observatories. Seven pioneers of astrophysics can be presented: William Huggins (1824–1910), and Joseph Norman Lockyer (1836–1920) near London, Angelo Secchi (1818–1878) in Rome, Pierre Jules César Janssen (1824–1907) in Paris, Nikolaus von Konkoly (1842–1916) in O'Gyalla, Hungary, and in addition the two early professional astrophysicists in German countries: As the cradle of astrophysics, the Observatory Bothkamp near Kiel (1869–1914) in the middle of a lake (IAU code 603, IAU List OAH: id 146) should be mentioned. Hermann Carl Vogel (1841–1907), scholar of Johann Karl Friedrich Zöllner (1834–1882), Professor for Astrophysics in Leipzig, started his career here, before he became director of the first *Astrophysical Observatory Potsdam*, Telegraphenberg (1874), 1882 to 1907.

1.1.2 New Instrumentation for Astrophysics

This new field of astrophysics caused, and was caused by, new instrumentation: spectrographs and object lens prisms, instruments for astrophotography (portrait lenses, astrographs), photometers for measuring the brightness of stars (starting with the visual Zöllner comparison photometer in 1860, later photographic and photoelectric photometry), solar physics instruments (since 1868, photoheliograph, spectroheliograph, later coronograph), and laboratory equipment for analysing the photographic plates and spectra (cf. blink comparator, spectrum comparator, measuring devices for getting the stellar magnitudes like iris diaphragm photometer, etc.).

1 Cf. the development of spectroscopy and spectral analysis: Wolfschmidt: Kosmochemie, 2022, p. 82–108.

2 *„Es kann nicht zweifelhaft geblieben sein, dass die Spectralanalyse hierbei die erste Stelle einnimmt und auch für absehbare Zeiten behalten wird, um so mehr, als durch die letzterwähnten Forschungen der Beobachtungs-Thätigkeit ein neues Feld eröffnet worden ist, welches in der Zukunft noch eine reiche Ausbeute verspricht.“* Scheiner, Julius: Die Spectralanalyse der Gestirne, 1890, p. 35.

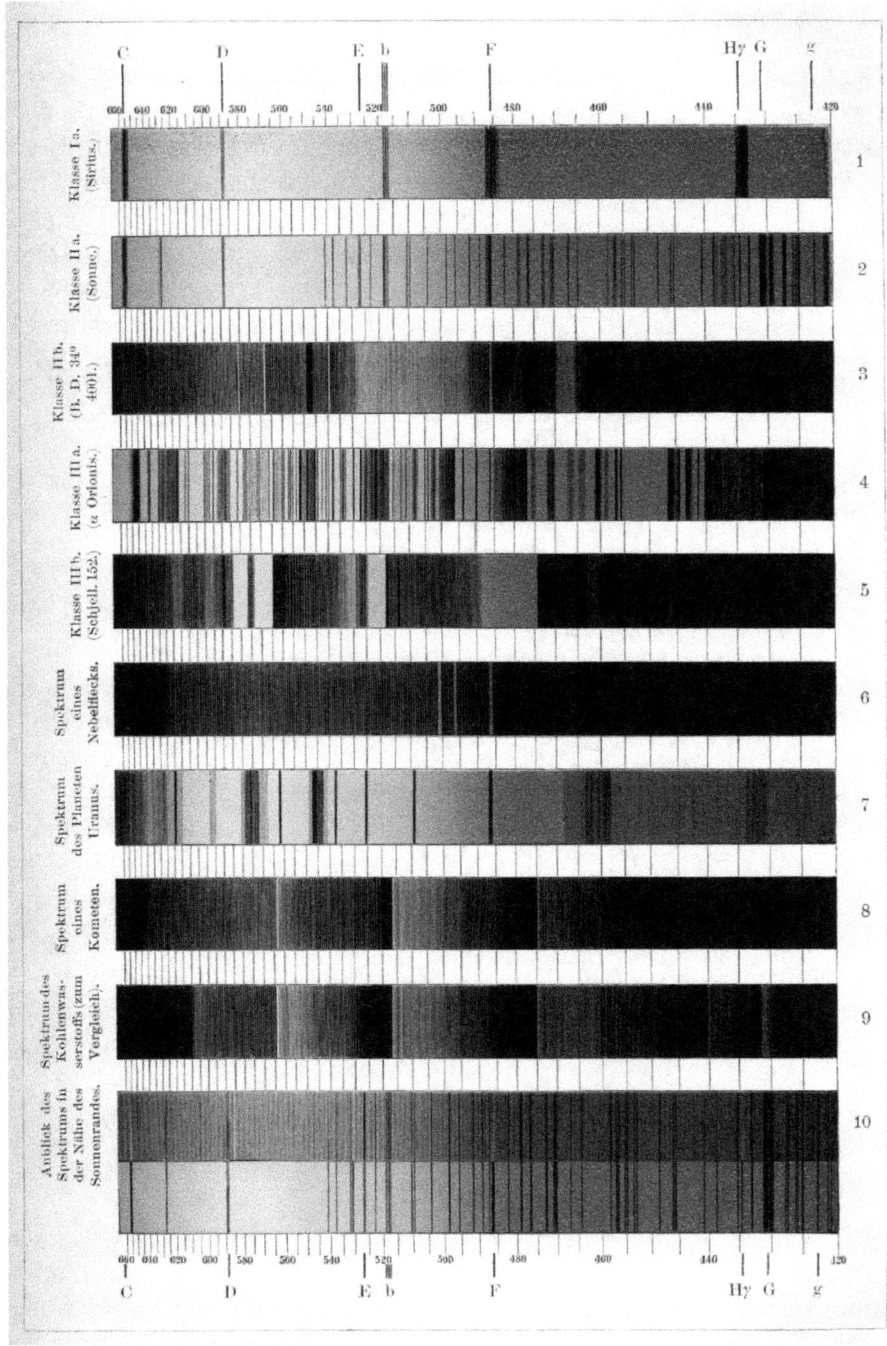

Figure 1.2:
Spectral classification with five classes, spectra of comets, Uranus, nebulae, and solar spectrum – Spectral table (1898)

Meyer, Max: Das Weltgebäude (1898), after p. 52.

1.1.3 Centers of Astrophysics until around 1900

In the last quarter of the 19th century only a few centres of astrophysics existed in the world.[3] Besides the Astrophysical Observatory Potsdam, where astrophysics was born, one should mention Göttingen, Heidelberg, Bonn, Bamberg, and Hamburg in Germany, then two observatories in Hungary, Italy (Collegio Romano Rome in the 1870s), England (Greenwich Astrophysical Department, 1873), France (Meudon Solar Observatory 1876), Russia (Pulkovo Astrophysical Department, 1882), and South America (La Plata, Argentina, 1886), also in the United States (Harvard College Observatory, Cambridge, Mass., (*1839), astrophysics with Edward Charles Pickering (1846–1919), director of HCO from 1877 to 1919, Lick, Mt. Hamilton, 1888, Yerkes, Wisconsin, 1897, Mount Wilson (1905), and India (Kodaikanal Solar Observatory, 1899 (since 1972, field station of the Indian Institute of Astrophysics in Bangalore).

Karl Schwarzschild was to play a special role in this phase of blossoming astrophysics.

1.1.4 New Observatory Layout – Astronomy Park

Around 1900, in addition, a revolution in observatory layout took place. This changed the architecture of the observatory building to the idea of an "Astronomy Park" with many different buildings and domes, as in the case of Nice Observatory (1879, id 225), EAO Kazan, Russia (1901, id 158), La Plata (1885, id 122), US Naval Observatory Washington D.C., 1887, id 116), Remeis Observatory Bamberg (1889, id 143), Uccle Observatory Bruxelles (1891, id 153), Heidelberg-Königstuhl (1896, id 141), and in Hamburg-Bergedorf (1906/12, id 92). The idea of an astronomy park observatory is performed in Hamburg in a perfect and consistent manner with a strict separation of observatory domes on one side, and the main building with the library, administration and offices, the workshop, and the residential buildings on the other side.

1.2 Biographical Notes on Karl Schwarzschild (1873–1916) – Places of Activity

Karl Schwarzschild (1873–1916) was born as the eldest of seven children in a German-Jewish upper middle-class family in Frankfurt am Main, where the

3 The *id* numbers of observatories refer to the IAU List (database) *Outstanding Astronomical Heritage* (OAH), `https://web.astronomicalheritage.net/heritage/outstanding-astronomical-heritage`.

Figure 1.3:
Karl Schwarzschild (1873–1916), (photo 1900)
Observatoire de Strasbourg, Refracting telescope

(Photo: Göttingen SUB Archive, "Cod. Ms. K. Schwarzschild – 23: 1,2").
(Top right: CC3, Ji-Elle, 2010), bottom: CC3, Pethrus, 2010)

proportion of Jews was 10% – the highest in Germany. He was the eldest son of his parents, Henrietta Ottilie Sabel (1852–1922) and Moses Martin Schwarzschild (1837–1916), a successful businessman. Karl attended the *Philanthropin*,[4] the primary school of the Israelite community. Then, he visited the *Städtisches Gymnasium* (Municipal Latin School, founded in 1519 during the Reformation movement) Frankfurt, a "Humanistic Gymnasium" (High School with Greek, Latin, and exact sciences), where he graduated in 1891. Karl and his friend Paul Epstein (1871–1939), the later mathematician, built a telescope, and they occupied themselves with mathematical and astronomical problems. At the age of 16, Schwarzschild published an article in *Astronomische Nachrichten* (1890) on determining the orbits of double stars using three measurements.

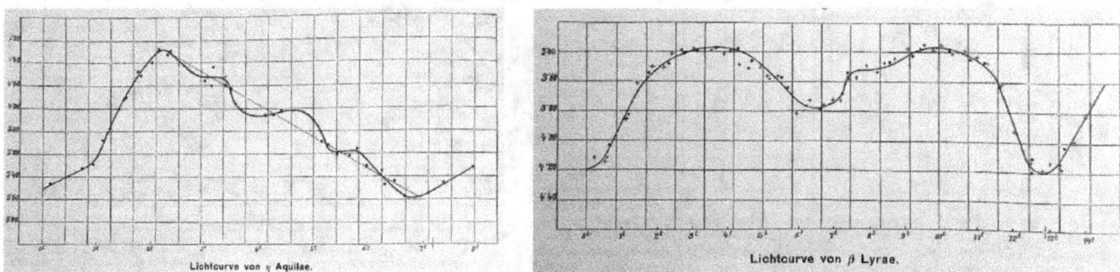

Figure 1.4:
Photometrische Beobachtungen:
Light curves of the eclipsing binaries η Aquilae (δ Cephei-Star) and β Lyrae

(Schwarzschild (1900a), VII. Capitel, C 123, Fig. 5+6)

Karl Schwarzschild[5] began his physics degree studies with Ernst Becker (1843–1912) in Straßburg/Strasbourg University. His first work concerned the observation of variable stars, Schwarzschild (1892).

He obtained his doctorate in Munich under Hugo von Seeliger (1849–1924) in 1896 with a mathematical thesis on Poincaré. During his time in Munich, from 1899 to 1901, he began to work on stellar statistics, an important field of research of Hugo von Seeliger.[6]

4 Johann Bernhard Basedow (1724–1790), founder of the philanthropic movement, improved the school system in line with Rousseau's natural educational concept. In accordance with the philanthropist's motto *For Enlightenment and Humanity*, the following content played a role alongside religious tolerance: modern languages, exact sciences, vividness and playful elements in lessons, patriotic education and the principle of equality.

5 Voigt (1989), p. 12–17. Bruggencate 1955, p. 232–239.

6 Schwarzschild: Über die Eigenbewegung der Fixsterne, (1907a), p. 614.

Figure 1.5:
Hugo von Seeliger (1849–1924), director 1882 to 1924.
Top right: 28.5 cm-Refractor, Fraunhofer, Merz of Munich (1835),
Bottom: Observatory Munich-Bogenhausen (1817), Refractor Dome

(Top left: © University Observatory Munich, Top right: CC3, Michael Florian Schönitzer,
Bottom: © University Observatory Munich)

Figure 1.6:
Top: Kuffner Observatory in Vienna-Ottakring, Bottom: Heliometer

(Top: CC3, Adolf Riess, Bottom: CC3, Tsui)

In the Kuffner Observatory in Vienna, founded in 1887, he created the new field of *photographic photometry* (1897 to 1899). He habilitated with this topic (Schwarzschild's law of blackening) at the University of Munich in 1899.[7] Then, he became a private docent; his lecture topics were nautical astronomy, astronomical optics and mathematical interpolation. In Munich Observatory, he devoted himself also to determining the orbits of spectroscopic binaries.[8] In addition, Schwarzschild published a paper on Arrhenius' theory of comet tails, a work that triggered a number of controversial discussions: He shows that the ion tails of comets cannot be explained by the radiation pressure of the Sun, as was assumed at the time. And this was correct; we know today that it is the solar wind.

Hugo von Seeliger proposed Schwarzschild for a professorship in Göttingen. This university was famous as a mathematical-physical centre thanks to Felix Klein (1849–1925), David Hilbert (1862–1943), Hermann Minkowski (1864–1909), Carl David Tolmé Runge (1856–1927), and Ludwig Prandtl (1875–1953). The first appointment list (26.7.1901) contained Seeliger, Max Wolf, and Schwarzschild, but he was rejected because he was too young (only 27 years)! Karl Hermann Struve of Königsberg was asked concerning the directorship (1895), but he declined – also the next astronomer Julius Franz (1847–1913), director in Breslau Observatory since 1897. Finally, Schwarzschild was called as Associate Professor of Astronomy and Director of Göttingen Observatory in 1901 – he was the youngest professor in Germany![9] His exceptional qualifications were quickly recognised, and he already became a Full Professor in 1902. In addition to the director, the observatory had only Leopold Ambronn (1854–1930) as observing astronomer, observing astronomer, Bruno Meyermann (1876–1963) as an assistant, as well as a warden, and an extra-budgetary auxiliary computer.

The instrumentation of the observatory was not particularly good at the time – quite outdated. Apart from a few smaller devices, there were essentially two instruments, the meridian circle, made by Reichenbach of Munich, from the Gauß era, and the large heliometer (15 cm aperture), made by Repsold of Hamburg, from Schur's time. In 1904, the observatory received a 19-cm-refractor (7 Paris inch), made by Reinfelder & Hertel of Munich, as a gift from Dr Anton Schobloch in Dresden; a separate building with a 4.5 m dome was erected in the observatory garden for this purpose. For his photographic work, Schwarzschild acquired a Zeiss Tessar lens in 1903.

7 Schwarzschild (1900a), p. C3–C135.
8 Schwarzschild (1900b), S. 65–74.
9 It is interesting, that the matter of his Jewish confession had not been discussed.

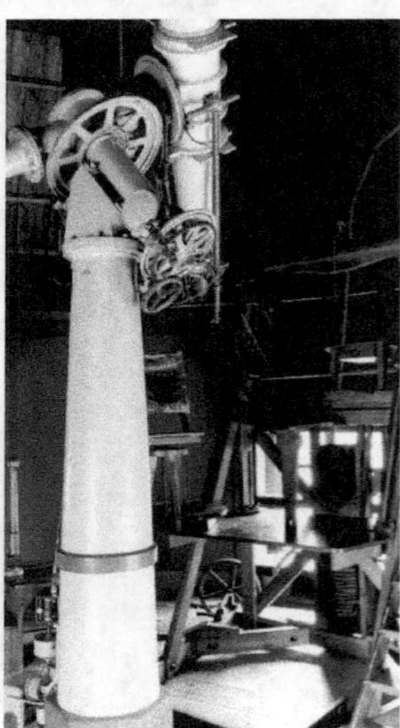

Figure 1.7:
Göttingen Observatory, the staff, and the 15-cm-refractor (1888)

(Top: Photos: © Gudrun Wolfschmidt, Bottom: Archiv Institut für Astrophysik Göttingen (IAG) und SUB Göttingen (Cod. Ms. K. Schwarzschild 23: 1,15)

As director of the Observatory in Göttingen in 1901, he devoted himself to theoretical optics, instrument development, and photographic photometry. Stimulated by the solar eclipse in Algeria (1905), he began studying solar physics, which in 1906 led him to the theory of stellar atmospheres. Karl Schwarzschild married Else Posenbach (†1950), daughter of a surgeon at the University of Göttingen, on 22 October 1909; they had three children: Agathe (Thornton) (1910–2006), Martin (1912–1997),[10] Alfred (1914–1944).

During the appointment negotiations as professor and director of the Astrophysical Observatory Potsdam in 1907/08, there were fierce disputes between the representative of classical astronomy, Arthur von Auwers (1838–1915), and the more progressive direction of the appointment committee, astronomer Karl Hermann von Struve (1854–1920), 1904 to 1919 director of Berlin Observatory, physicist Emil Warburg (1846–1931), geodesist Robert Helmert (1843–1917), who wanted to strengthen the new astrophysics.[11] The 28-year-old Schwarzschild was in second place on the list after Hugo von Seeliger.

> *"He* [Schwarzschild] *is equally at home in astronomy and astrophysics, as well as in the border areas of physics, masters mathematical analysis in a perfect manner and, in addition to comprehensive knowledge, possesses unusual working power."*[12]

But Auwers initially rejected Karl Schwarzschild at first, mainly because he was too versatile and and neglected classical astronomy in favour of physics and mathematics. Only after two years of negotiations, Auwers was outvoted.[13]

10 Martin Schwarzschild was already during the Gymnasium also very much interested in mathematics and astronomy: The retired Carl Runge became his tutor: *"Carl Runge, for example, one of the real close friends from that time, permitted me for years, when I was towards the last years of high school, to spend one afternoon a week with him. He got me into mathematics and physics well ahead of what I could get in high school." "My mother took me a couple of times to the then director of the observatory, Professor Hans Kienle, who later became my main teacher in astronomy – took me to him already when I was a high school student.* [dots] *By that time I had put together a little telescope and an apparatus to photograph stars with a small box camera, in what free time the school left me. And after a while, Professor Kienle had seen enough of that, that he loaned me from the observatory an old four-inch* [Fraunhofer] *telescope."* Schwarzschild, Martin – AIP Interview Spencer Weart: `https://www.aip.org/history-programs/niels-bohr-library/oral-histories/4870-1`, accessed 12 December 2024.

11 Treder (1974), p. 13–19, here p. 14.

12 Gutachten (Expert opinion) der Akademie über Karl Schwarzschild, Dezember 1907, 2nd expert opinion 1909. Archive material in the Academy of Sciences in Berlin, II: VIa Band 12,13 (Berufung Schwarzschilds nach Potsdam), II: IIIa Band 18 (Wahl Schwarzschilds zum Akademiemitglied).

13 Auwers' preferred candidate would have been Gustav Müller (1851–1925), who then became his successor after Schwarzschild's unexpectedly early death in 1916. Although

After Julius Scheiner's (1858–1913) death in 1913, professor of astrophysics in Potsdam, it was finally agreed – two years later – in December 1915, to have the *„physical parts of astronomy"* represented by two professorships at the University of Berlin:[14]

- a – first – chair for *theoretical astrophysics*, which Karl Schwarzschild was awarded on 5 February 1916, and
- an associate professorship for experimental or *observational astrophysics*, to which the photometrician Paul Guthnick (1879–1947) was appointed.

As director of the Astrophysical Observatory Potsdam, 1909 to 1916, Schwarzschild was particularly concerned with the General Theory of Relativity that was just emerging, besides solar physics, spectroscopy,[15] radiation and quantum physics, stellar statistics, and stellar dynamics. Karl Schwarzschild and his important work on photography, photometry and especially on the foundation of theoretical astrophysics marked another highlight in the development of this observatory. Schwarzschild was convinced that the interplay with other areas of exact sciences was of great importance for the development of astrophysics. As a result, astrophysical observatories were established precisely where there were well-known physics or chemistry institutes.

In 1914, at the age of only 41, Karl Schwarzschild volunteered to join the German army in World War I[16] (cf. chapter of Adriaan Raap, 4, 131). He served on the northern and eastern fronts and was promoted to lieutenant of artillery. He died in 1916 of a then incurable skin disease *pemphigus*.[17]

Müller was a representative of visual photometry, his achievements and creativity could not be remotely compared to Schwarzschild. In addition, the era of photographic photometry with Schwarzschild and, from 1913, even the era of photoelectric photometry with the pioneers Paul Guthnick (1879–1947) and Hans Rosenberg (1879–1940) had long since dawned.

14 Herrmann 1977, here p. 42.

15 Schwarzschild: [β Aurigae], (1900c). Schwarzschild: [Nova Geminorum], (1912b).

16 Martin Schwarzschild, why his farther decided to go to the army: *"According to my mother, it was entirely one thing that determined my father, and that was the conviction that German Jews could not expect to overcome the anti-Semitism if they did not lean over backwards in doing their national duty. It was that reason and that reason only for which he did go into the army."* Schwarzschild, Martin – AIP Interview Spencer Weart: `https://www.aip.org/history-programs/niels-bohr-library/oral-histories/4870-1`, accessed December 12, 2024.

17 His papers (*Nachlaß*) in the Göttingen State and University Library comprises around 7,000 letters from around 700 correspondents, edited in 1992 by Hans-Heinrich Voigt (1921–2017). Cf. Voigt 2011.

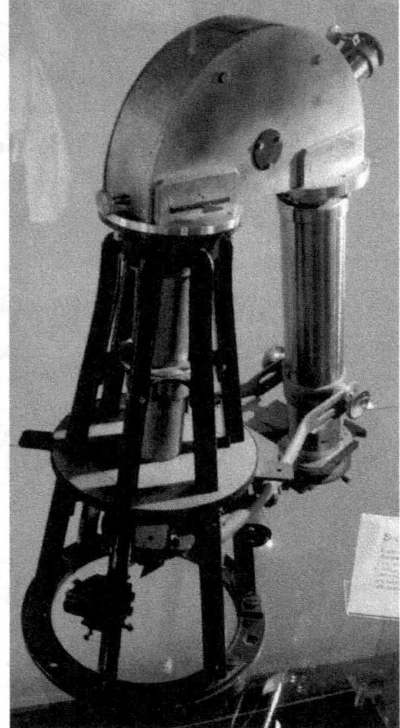

Figure 1.8:
Astrophysical Observatory Potsdam, Large Steinheil-Repsold Refractor, 3-Prism-Spectrograph

(Photos: Gudrun Wolfschmidt)

1.3 Biographical notes on Ejnar Hertzsprung (1873–1967) – places of activity

The parents of Ejnar Hertzsprung (1873–1967) were Severin Carl Ludwig (1839–1893) and Henriette Hertzsprung (1839–1915); they lived in Copenhagen.[18] His father had a doctorate in astronomy, but worked in an insurance company for lack of employment in the astronomical field. Ejnar Hertzsprung studied chemical engineering at the Polytechnical University Copenhagen (1893/98), and worked then in a chemical factory St. Petersburg.

In Leipzig, Hertzsprung started a PhD project about photochemistry with Wilhelm Ostwald (1853–1932) as supervisor (1901/02). In 1902, he returned to the University of Copenhagen, and worked at Urania Observatory in Frederiksberg.

1.3.1 Ejnar Hertzsprung and his Colour-Brightness Diagramme

In 1905 and 1907, Hertzsprung published now classic articles *Zur Strahlung der Sterne* (On the Radiation of Stars) about his attempts to measure the luminosity of stars on the basis of their spectra. Karl Schwarzschild became aware of the young talent and invited him to Göttingen in 1908. He was convinced of Hertzsprung's abilities, and offered him an assistant professor position at Göttingen University in 1909. Their cooperation was short, because Schwarzschild got the call to Berlin University, and as director of Potsdam Astrophysical Observatory. He succeeded again to find a position for Hertzsprung. After Schwarzschild's death in 1916, and due to the problems as a foreigner during the World War I, he went to the Leiden University Observatory, Netherlands, in 1919. He became director there from 1935 to 1944, and returned to Denmark a year before his death, living near the Brorfelde Observatory (OAH id 241) south of Copenhagen.

In 1905, he defined the absolute brightness as a measure of the luminosity of a star. He also recognized that the later so-called "giant stars" and "dwarf stars" can occur with the same surface temperature. Inspired by this idea, he designed already in 1905 a colour-brightness diagramme – later crucial for stellar development.

Karl Schwarzschild met in 1910 during his trip through the USA Henry Norris Russell (1877–1957), who was not aware of Hertzsprung's similar activities. Russell then presented his diagram in 1912. Later, this was called Hertzsprung-

18 Concerning his biography see Herrmann 1994/2011.

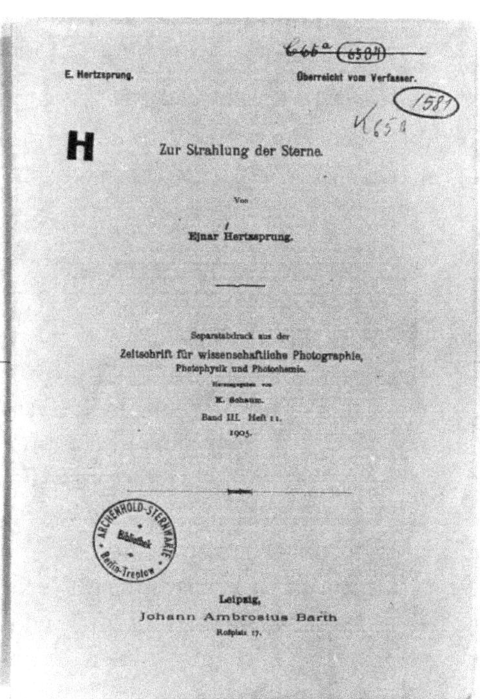

E. Hertzsprung. Überreicht vom Verfasser.

H

Zur Strahlung der Sterne.

Von

Ejnar Hertzsprung.

Separatabdruck aus der

Zeitschrift für wissenschaftliche Photographie, Photophysik und Photochemie.

Herausgegeben von

K. Schaum.

Band III. Heft 11.

1905.

Leipzig,

Johann Ambrosius Barth

Roßplatz 17.

Figure 1.9:
Top: Ejnar Hertzsprung (1873–1967): Zur Strahlung der Sterne (1905/07), Bottom: Astrophysical Observatory Potsdam, Old Leiden Observatory (**scienzapertutti.infn.it**, CC, photo: Gudrun Wolfschmidt, CC)

Russell Diagramme (HRD), developed independently by these two researchers. See the detailed development, p. 41.

50 years after Kirchhoff and Bunsen, the development of astrophysics reached a peak with Schwarzschild and Hertzsprung. Modern theoretical astrophysics began through the inclusion of important physical topics in astronomy.

1.4 Transition from Visual to Photographic Photometry

Karl Schwarzschild is regarded as the leading inventor of photographic photometry. Around 1900, this innovative method made it possible to measure the photographic (chemical) effect of the light energy objectively – instead of using visual photometry as a subjective method. Due to many pioneering works in the field of astrophotography, there was great interest in determining the brightness of stars on photographic plates. The question was whether the diameter or the blackening was more suitable for this purpose. The normal blackening law of Robert Wilhelm Bunsen and Henry Enfield Roscoe (1833–1915) from 1862 stated that the degree of blackening S increases proportionally with the illumination intensity I and exposure time t

$$S = I \cdot t$$

This essentially applied to exposure times between approximately 1/1000 and 1 second. Schwarzschild found that this not true for the long exposures that were typical in astronomy.

Schwarzschild first investigated this question with extrafocal photography (Schwarzschild, 1900c), later also with focal photography, and his invention,[19] the *Schraffierkassette* (hatching cassette), also known as a *Wackelkamera* (wiggle camera) with a 4-cm-Zeiss-Tessar (1:10), later built by Askania of Berlin.[20] This special device moved the photographic plate back and forth during the exposure in such a way that the light of each individual star was smeared onto a small square of $0.25 \times 0.25\,mm^2$ within three minutes. The blackening on the photographic plate could now be measured precisely. These experiments finally led to the introduction of the *Schwarzschild exponent* p = 0.73–0.95. This de-

19 This new instrument was developed with the help of Bruno Meyermann (1876–1963). Schwarzschild & Meyermann: (1906), p. 277–282. Schwarzschild (1907b), p. 137–140.

20 Bamberg, Carl: Katalog Astro 40, Berlin 1924, p. 41, Fig. 27. Deutsches Museum, Sondersammlung, Firmenschriften 271 Askania.

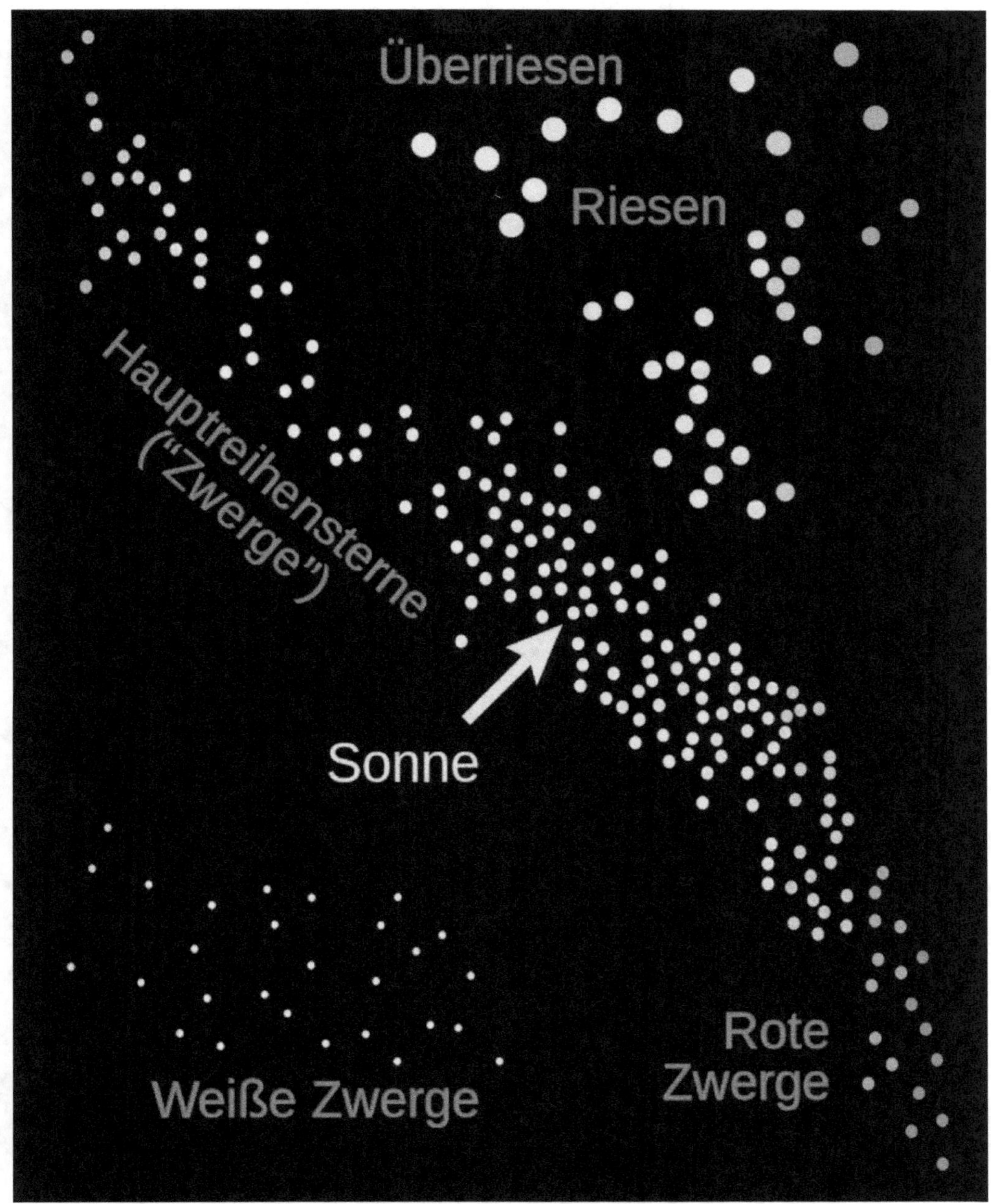

Figure 1.10:
HRD (Riesen- und Zwergsterne)

(CC3, `HR-sparse-de.svg`, User:Rursus)

scribes the relationship between the amount of light and blackening during the exposure time, i. e.

$$S = I \cdot t^p$$

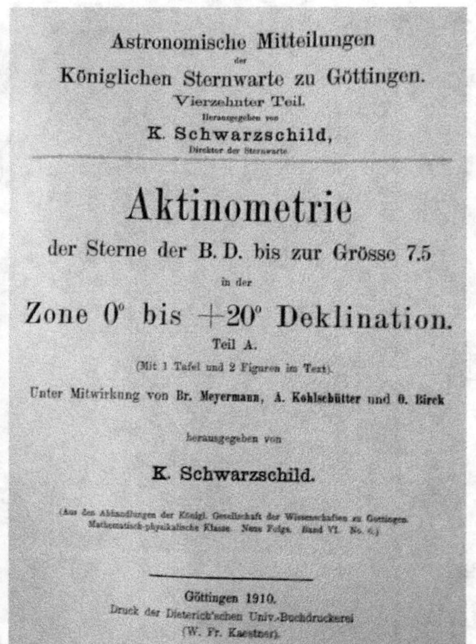

Astronomische Mitteilungen
der
Königlichen Sternwarte zu Göttingen.
Vierzehnter Teil.
Herausgegeben von
K. Schwarzschild,
Direktor der Sternwarte

Aktinometrie
der Sterne der B. D. bis zur Grösse 7.5
in der
Zone 0° bis +20° Deklination.
Teil A.
(Mit 1 Tafel und 2 Figuren im Text).
Unter Mitwirkung von Br. Meyermann, A. Kohlschütter und O. Birck
herausgegeben von
K. Schwarzschild.

(Aus den Abhandlungen der Königl. Gesellschaft der Wissenschaften zu Göttingen. Mathematisch-physikalische Klasse. Neue Folge. Band VI. No. 6.)

Göttingen 1910.
Druck der Dieterich'schen Univ.-Buchdruckerei
(W. Fr. Kaestner).

Figure 1.11:
Left: Schwarzschild's *Schraffierkassette* or *Wackelkamera*
(hatching cassette or wiggle camera)
Right: Schwarzschild's Göttingen Aktinometrie (1910/12)

(Schwarzschild, (1907b), Schwarzschild, Teil A, 1910a)

Schwarzschild's photographic brightness catalogue, the *Göttingen Aktinometrie*[21] with 3500 stars up to magnitude 7.5^m, was created on this basis in 1910. However, this early photographic brightness catalogue was never completed. The complete work should have included 14,199 stars as a photographic counterpart to the Potsdam Durchmusterung (survey) with the visual magnitudes. Schwarzschild started to compare his photographic (blue) brightnesses with the Potsdam visual (yellow-green) brightnesses. This led to the introduction of the concept of *Farbtönung* (today called 'colour index'), which opened the way to

21 Schwarzschild Teil A & Teil B (1910, 1912).

multicolour photometry. Together with Hertzsprung, this led to the realisation of the existence of red giant stars, and the Hertzsprung gap. The plan was to provide a large amount of material for determining the colour and temperature of the fixed stars. There were even more far-reaching plans, namely the photographic recording of all the stars in the *Bonner Durchmusterung*. Due to these activities, Karl Schwarzschild can be regarded as pioneer of observational astrophysics.

From the 1920s onwards, new instruments such as *iris diaphragm photometers* made it possible to simultaneously measure the diameter and blackening on photographic plates, and to make precise brightness measurements.[22] However, the lasting significance of photographic photometry is based on the statistical recording and permanent documentation of brightness in large areas of the sky. Only then, it was possible to systematically search for variable stars.[23]

1.5 Theoretical Optics

A large double refractor was purchased for the Astrophysical Observatory Potsdam in 1899. The optics came from Steinheil of Munich, and consisted of an 80-cm-refractor for photography and a second 50-cm-refractor for visual observation. Due to the unsatisfactory quality, Schwarzschild began negotiations in 1910 with the – at that time still relatively unknown – optician Bernhard Schmidt (1879–1935) about possible improvements to the large refractor. However, the Steinheil company tried to prevent this because Steinheil saw its reputation jeopardised; he intervened in 1912 with a letter of petition to Schwarzschild.[24] Nevertheless, Schwarzschild prevailed, and Schmidt successfully corrected the 50-cm-lens in 1912.

With regard to the correction of the 80-cm-lens, Schwarzschild wrote to the Minister of Clerical and Educational Affairs in 1913: *"The honourable undersigned requests Your Excellency to entrust the correction to Mr Schmidt. The undersigned is convinced that Mr Schmidt that Mr Schmidt will carry out the correction of the lens better and faster than the Steinheil company. Mr Schmidt*

22 Wolfschmidt: Entwicklung der astronomischen Photometrie, 1989, p. 227–268, p. 271–272.

23 The next step was the introduction of photoelectric photometry in 1913 by Paul Guthnick in Potsdam and Hans Rosenberg in Tübingen; for the world's first photoelectric photometer a potassium photocell was used. Wolfschmidt: Hans Rosenberg, 2014, p. 280–311. Photography was important until the 1980s, then, in the 1990s, the era of modern digital CCD images started.

24 Steinheil, Rudolf: Brief an Karl Schwarzschild vom 1. Feb. und 7. März 1912. Schriftwechsel mit Schmidt in Mittweida, 1910–1916. Archiv der Akademie der Wissenschaften in Berlin, Astro Obs 31.

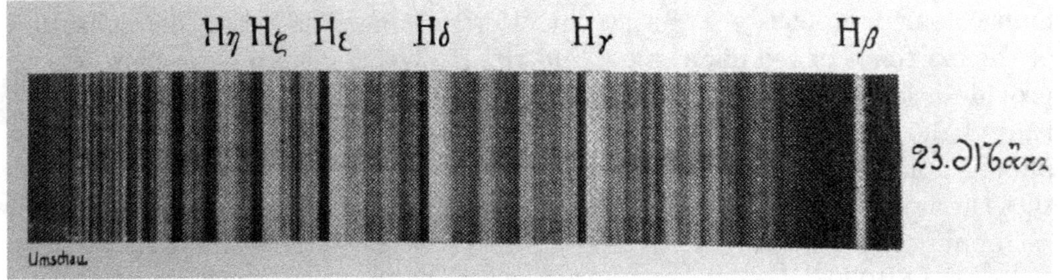

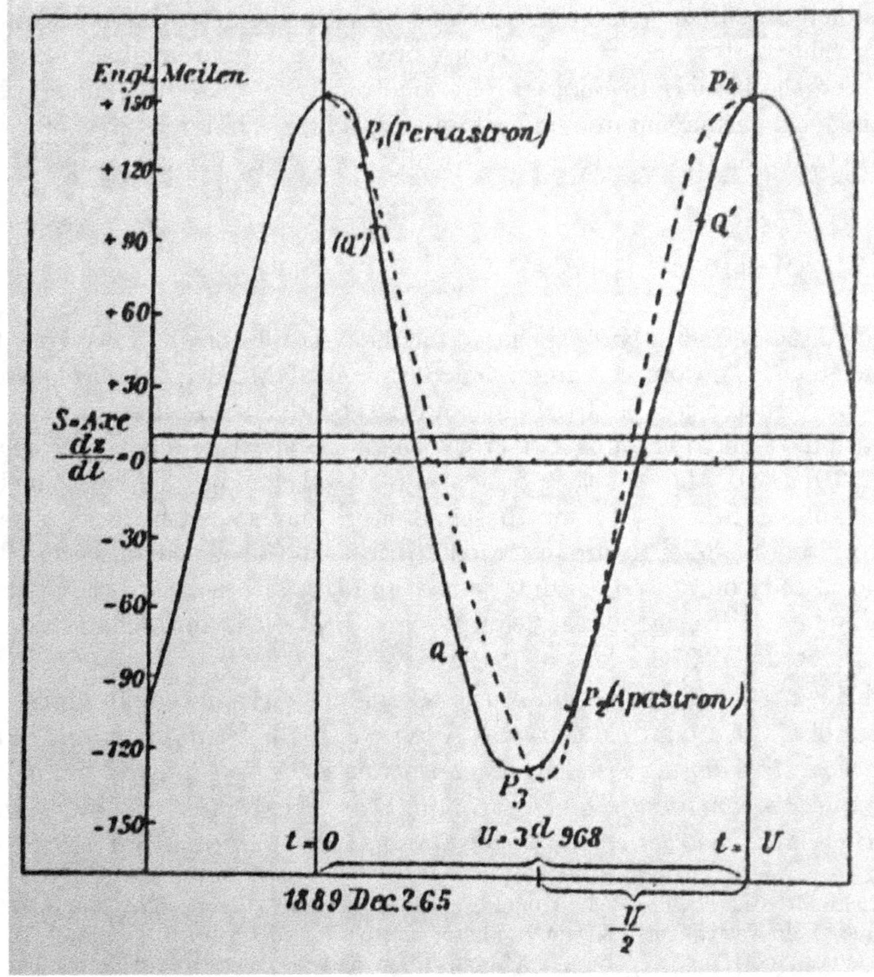

Figure 1.12:
Schwarzschild's spectroscopic observations:
Top: Spectrum of Nova Geminorum No. 2, observed at APO Potsdam
Below: Orbit determination of the spectroscopic binary β Aurigae

(Schwarzschild (1912b), 637; Schwarzschild (1900c), 71)

is the greater artist, and since it is a matter of perfecting the largest German objective lens as far as possible, the undersigned believes that all other considerations must take a back seat."[25] But in this case, Steinheil succeeded in 1913 in preventing the correction the 80-cm-objective lens; the result was not convincing.

With Karl Schwarzschild, interest shifted to theoretical optics. At that time, people were interested in the problem of how a large celestial field up to 15° could be imaged without distortion. With reflecting telescopes, only a small field of view up to about 15′ is available. A mirror with a parabolically ground surface a distortion-free (coma-free) image is only produced when the rays are parallel to the optical axis. Inclined rays produce asymmetrical, comet-like images. This image aberration is therefore called coma.

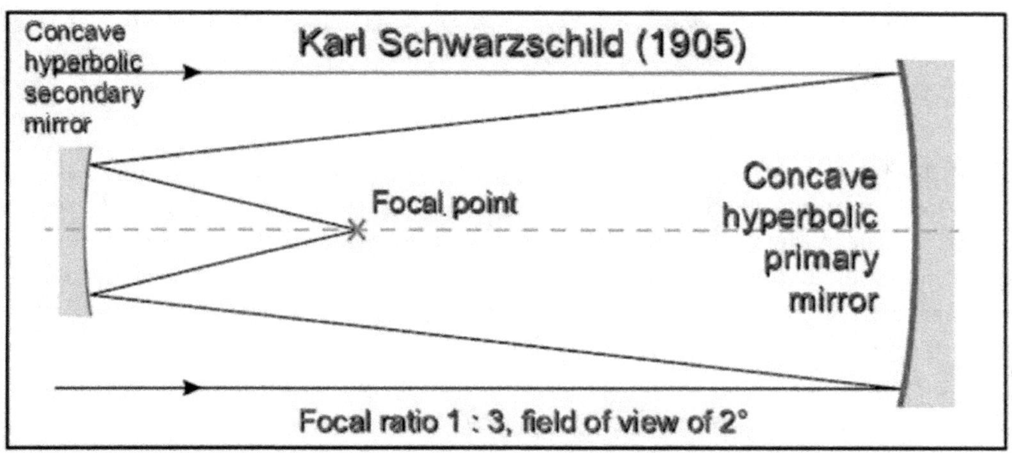

Figure 1.13:
Schwarzschild's reflecting telescope with a field of view of 2° (1905)
Concave hyperbolic primary and secondary mirror, focal ratio 1 : 3

Schwarzschild first thoroughly investigated all imaging aberrations of optical systems: spherical aberration, coma, astigmatism, field curvature, distortion. Subsequently he developed a theory of the reflecting telescope. He proposed

25 Schwarzschild, Karl: Brief an den Minister der geistlichen und Unterrichts-Angelegenheiten vom 7. Juli 1913. Schriftwechsel mit Schmidt in Mittweida, 1910–1916. Archiv der Akademie der Wissenschaften in Berlin, Astro Obs 31.

the combination of two mirrors with a concave hyperbolic primary mirror and a concave elliptical secondary mirror. These were free from spherical aberration and coma, and only exhibited the image aberration of astigmatism. An aperture ratio of mirror diameter to focal length of 1:3 produced a field of view of 3°.

However, the system had a number of disadvantages, such as the long overall length and the shadowing caused by the secondary mirror. The manufacture and testing of such aspherical mirror systems was particularly problematic. Inspired by Schwarzschild, Henri Jacques Chrétien (1870–1956) also constructed a telescope with a larger field of view in 1922, which exhibited the same and similar problems.

Karl Schwarzschild held Bernhard Schmidt's work in high esteem. In July 1913, he wrote to the Minister: *"The confidence of the undersigned in the quality of Mr Schmidt's work is based* [...] *on the personal knowledge of his extraordinarily clear insight into the mathematical-physical principles of telescope optics, which he was able to acquire essentially as an autodidact."*[26]

Schwarzschild was proved right. In 1930, Bernhard Schmidt achieved maximum field of view with optimum image quality without spherical aberration, coma, and astigmatism with the type of telescope named after him. To avoid the coma, Schmidt used a spherically ground mirror (spherical mirror), which is easy to manufacture. However, this gave him an additional aberration, the spherical aberration. He removed this aberration by using a specially calculated aspherical correction plate, which he placed at the centre of curvature of the spherical mirror. Schmidt also developed an ingenious idea for the practical production of this correction plate, namely with the help of evacuation. The focal ratio of 1:1.75 and the large field of view of 15° with distortion-free imaging was impressive.[27]

1.6 Planning a Southern Observatory

Around the turn of the century, proposals and suggestions for a Southern Observatory came from Max Wolf (1863–1932), director of the Heidelberg Observatory, under the influence of his trip to America in 1893. Hermann Carl

26 Schwarzschild, Karl: Brief an den Minister der geistlichen und Unterrichts-Angelegenheiten vom 7. Juli 1913. Schriftwechsel mit Schmidt in Mittweida, 1910–1916. Archiv der Akademie der Wissenschaften in Berlin, Astro Obs 31.

27 The world's largest Schmidt telescope with a mirror diameter of 2 m, a free aperture of 1.34 m, and a focal length of 4 m is located at the Karl Schwarzschild Observatory in Tautenburg near Jena.

Vogel (1841–1907), Director of the Astrophysical Observatory in Potsdam Astrophysical Observatory, planned to establish a branch station on the Spanish Mediterranean coast, but Vogel died before the project could be realised.

Schwarzschild's interest in solar and stellar physics was expressed in 1907, when he and professors of physics, geo science, and mathematics submitted an application for the establishment of an observatory for solar physics and geophysics.[28] Four years later, the geophysical and astronomical institutes of the University of Göttingen submitted a petition to the Ministry of Culture for the establishment of a southern branch of geophysics, the Samoa Observatory in the Pacific Ocean.[29] This application was elaborated in more detail in 1908, and specifically astrophotographic instrumentation was proposed, including a 15 cm refractor, two Schwarzschild wiggle cameras, and special optics for astrophotography.

In 1907, the physical-astronomical institutes of Göttingen University submitted a petition to the Ministry of Culture to establish a southern branch:

> *"The Samoa Observatory works exclusively in the field of geophysics. It seems urgently desirable to us to also have a similar observatory for the purposes of astrophysics, because in Germany two things are unalterably lacking that such an observatory, located in the southern hemisphere, would have: good weather and visibility of the southern sky.*
> *Scientific tasks of the station.*
> *a. Photographic photometry of the fixed stars.*
> *At Göttingen Observatory, the photographic magnitudes of stars up to 7.5 are determined for the zone 0 degrees to 20 degrees declination. The accuracy of these determinations is unprecedented (2% probable error of the stellar magnitudes), and the statistical analysis of the results is of great importance for the knowledge of the constitution of the fixed stars. The continuation of this work for the southern sky would have to be the main purely astronomical task of the station.*
> *b. Measurement of solar radiation.*
> *The recently deceased American physicist Langley claimed that the sun's radiation is variable, and that it decreased by 10% in 1903,*

28 Schwarzschild, Karl; Voigt, Woldemar; Wiechert, Emil; Wagner, Hermann; Klein, Felix; Riecke, Eduard: Betrifft Einrichtung einer südlichen Filiale der Göttinger physikalisch-astronomischen Institute. 7. August 1905. An den Minister der geistlichen, Unterrichts- und Medicinalangelegenheiten. Universitäts-Sternwarte Göttingen.

29 Angenheister: Geschichte des Samoa-Observatoriums, 1974, p. 43–66.

Figure 1.14:
Left: Samoa Observatory of Göttingen University,
Right: Cuno Hoffmeister (1892–1968) Windhoek Observatory (1938)

(CC)

> *for example. This decrease in solar radiation should also be associated with a decrease in temperature on the Earth's surface, and Langley tried to prove this decrease in meteorological observations for various locations.* [...] *The independent examination of Langley's results at a completely different location on Earth seems to us to be of the greatest interest."*[30]

In 1910, after a round trip to the famous large American observatories, Karl Schwarzschild once again emphasised the need for a German outstation, firstly because of the weather conditions in Germany, and secondly in order to be able to observe the sky in southern latitudes. He set his sights on Windhoek in German South West Africa (today Namibia).[31] A site testing expedition to Tenerife was undertaken by the Potsdam astronomers astronomers Gustav

30 Schwarzschild, Karl: Betrifft Einrichtung eines Observatoriums der kgl. Gesellschaft der Wissenschaften zu Windhuk. 27 July 1908. University Observatory Göttingen.
31 Schwarzschild (1910b), p. 1531–1544.

Müller (1851–1925) and Erich Kron (1881–1917) as early as 1910.[32] However, Schwarzschild's endeavours remained unsuccessful. World War I put a stop to all plans.

Around 1910, there were already two bases of German physics and astronomy abroad: The Göttingen astronomer Bruno Meyermann (1876–1963) had been in Tsingtau (Qingdao) in China since 1908. In 1909, the German physicist Emil Bose (1874–1911) was appointed to the university in La Plata, Argentina; in 1911, Johannes Hartmann (1865–1936) followed him to the Argentinian National Observatory in La Plata.

It was not until Cuno Hoffmeister (1892–1968) succeeded in establishing an observatory in Windhoek in 1938.

1.7 Spectral Classification and Hertzsprung-Russell-Diagram (HRD)

Objective prisms are important for the classification of stellar spectra because it is easy to record many spectra simultaneously. Angelo Secchi (1818–1878) in Rome in 1863 was a pioneer in using an objective lens prism with his 24-cm refractor for classifying spectra.

Karl Schwarzschild in particular recognised the importance of the objective prism and used it, especially in the UV range, and to determine radial velocities. At Schwarzschild's suggestion, Hans Rosenberg (1879–1940) photographed the Pleiades with a UV prism, made by Carl Zeiss of Jena at the Göttingen Observatory. He analysed the strength of the H and K lines in comparison to the hydrogen lines (H_δ, H_ζ) as an indication for the spectral type, plotted as abscissa, with the photographic brightness as ordinate. This work, Rosenberg's habilitation (1901), on the relationship between brightness and spectral type in the open star cluster of the Pleiades, is the first colour-brightness diagramme.

Edward Charles Pickering (1846–1919), who created the basis for the classification of stellar spectra, gained great experience with objective prisms. Annie Jump Cannon (1863–1941) worked there from 1888 to 1935, one of the many female astronomers in 'Pickering's harem'.[33] She analysed more than 1100 southern sky stars on 5961 plates. She published her first classification in 1901. Cannon developed a *3rd Harvard Classification*; this finally resulted in the designation of the types as O, B, A, F, G, K and M. With this, Cannon

32 Müller, Gustav: Brief an Karl Schwarzschild vom 18.4.1910 aus Teneriffa. Niedersächsische Staats- und Universitätsbibliothek Göttingen, Handschriftenabteilung, Nachlaß Karl Schwarzschild, Briefe 535, G. Müller, p. 10.

33 Hoffleit: Pioneering Women in the Spectral Classification of Stars (2002), p. 370–398.

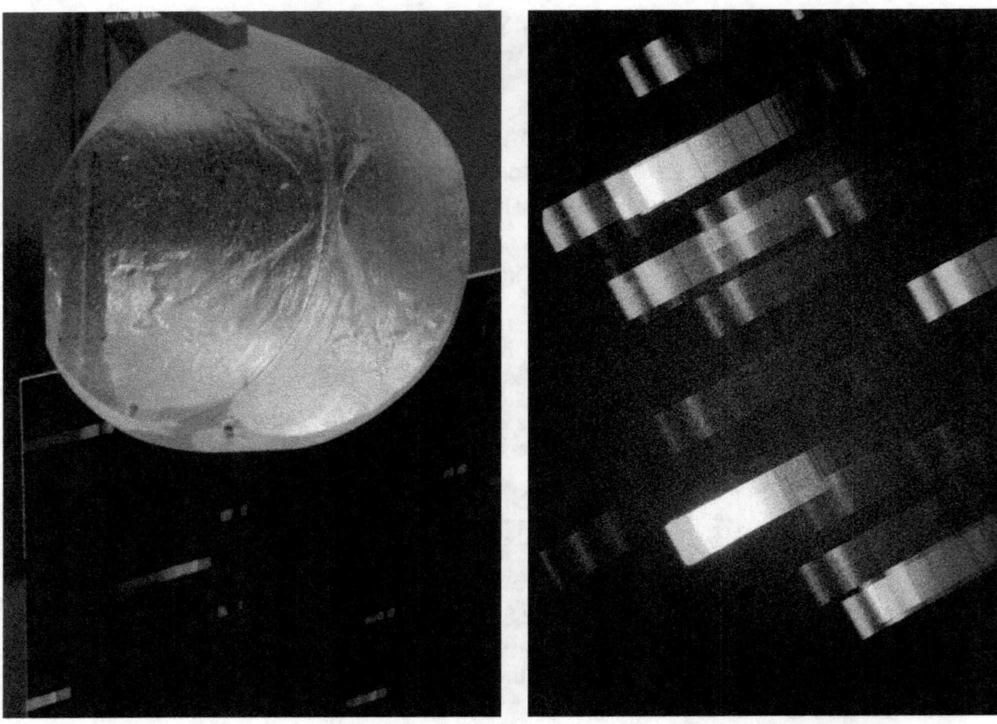

Figure 1.15:
Objective lens prism and objective prism spectra
(Photo: Gudrun Wolfschmidt)

compiled the extensive *Henry Draper Catalogue*, nine volumes (1918 to 1924). This gradually replaced the Vogel classification (blue, yellow, and red stars).

> *"In my opinion, the Draper classification is the best currently in existence. [...] The Draper classification represents the spectra as a function of a variable which can be described as the 'developmental stage'. Miss Maury already adds the line width as a second variable, which, according to Hertzsprung's investigations, is a very important physical criterion."*[34]

Cannon's spectral classification was a great achievement; in 1922 it was adopted by the International Astronomical Union (IAU) as an international standard.

34 Schlesinger (1911), p. 297–298.

There was a puzzling problem at that time: If all star types were ordered according to increasing proper motion, i. e. decreasing distance, then one gets a sequence from blue to red to yellow, which means that the yellow stars have the lowest luminosity. This sequence contradicted stellar evolution, in which a contraction and cooling from blue via yellow to red was assumed. This contradiction could not be resolved at the time.

William Henry Stanley Monck (1839–1915)[35] was interested in the distribution of stars in space; for this he needed distances and absolute magnitudes. From 1892 to 1898, Monck attempted in 12 papers to find a connection between the proper motion of 107 stars from the Bossert catalogue, and their spectral type from the *1st Harvard Catalogue* of 1890. Then, he attempted a brilliant explanation as early as 1894: He thought about the colours and spectra of the stars. He found a correlation between proper motion and spectral type: The proper motions are increasing (and distance and luminosity are decreasing) from the Blue stars or *Sirian stars* (spectral type B and A, 11 stars) to the Yellow stars or *Capellan stars* (spectral types F and G, 92 stars), like the Sun or α Centauri, then proper motions are decreasing from the Yellow to the Red stars or (*Arcturian stars* (spectral type K and M, 59 stars). The result was: There could be two classes of yellow stars, the one distant and luminous like Capella, the others close and about as bright as the Sun. With this finding, he effectively postulated the existence of »giant« and »dwarf« stars.[36] The c-stars are characterised by high luminosities and large distances, which is reflected in their small proper motion.[37] The idea of giant stars is based on the fact that only so much light can be emitted from cool stars if their surface area is very large, i. e. if they are giant stars with a very large surface – with diameters of 10 to 100 times the radius of the Sun. If he would have had more and better data, his discussion of proper motions and spectra might have led him to the relationship between luminosity and colour, later discovered by Hertzsprung and Russell.[38]

35 Monck, educated in Trinity College Dublin, was called as professor of moral philosophy in Trinity College until 1882, then in 1886, fellow of the Royal Astronomical Society and member of the Liverpool Astronomical Society. In 1891, he purchased a 7½-inch refractor, made by Alvan Clark. Cf. Elliott, 2007.

36 Monck: The Spectra and Colours of Stars, (1895).

37 Because the material of proper motions was still very small, they only had the misfortune of having only distant giant stars with small proper motions available among the red, faint stars, while among the yellow stars they had both giants and normal, faint dwarf stars, which together had a larger average proper motion.

38 De Vorkin: Stellar Evolution and the Origin of the Hertzsprung-Russell Diagram, 1984, p. 90–108.

These indications of different groups of stars with the same spectral type were in agreement with the results of Antonia C. Maury (1866–1952), who in 1897 found c-stars with sharp spectral lines on the one hand, and non-c-stars with 'normal', broad absorption lines of hydrogen, and weak, diffuse metal lines on the other hand. She considered these types of stars as parallel evolutionary paths.

Ejnar Hertzsprung (1873–1967) compiled his "diagramme" (he made tables, but not a graphic representation) from 1905 onwards.[39] He found that the c-stars were characterised by high luminosity and great distance, which was noticeable by their small proper motion. When Hertzsprung turned to this topic in 1905, he was fortunate that more values for the proper motions were available in order to investigate a connection with the spectra. In particular, the *2nd Harvard Classification* developed by Maury was the key to Hertzsprung's understanding that there are not only dwarf stars, i. e. 'main sequence stars', but also giants of high luminosity with the same spectral type, which are noticeable in the spectra as c-stars with sharp lines. However, he had not yet used the term 'giants'.

The c-stars are therefore the »giant stars«, as they were called by Antonia Maury. But she was treated by Pickering in an unfair way. He did not like that she was not following his orders, and using the existing classification. Instead, she introduced by carefully examining the spectra, additional features like the a, b, and c divisions (indicating if the spectral lines are relatively wide (normal), hazy, or exceptionally sharp). Pickering assumed that these are only different quality features of the photographic plates, but not intrinsic stellar characteristics.

Karl Schwarzschild was fascinated by Maury's work, and understood the enormous significance of this discovery – especially because he was familiar with the work of Rosenberg and Hertzsprung in this same direction.

> *"It is in itself most curious and unforeseeable by any theory of stellar evolution that these giants should lie so scattered among the ordinary stars. Once again, the world proves to be the noblest work of art, never arbitrary, and yet always surprising. But even more promising is the view that opens up from this final point of our advance."*[40]

During his visit by Karl Schwarzschild to America in 1910, he tried in vain to convince Pickering of the scientific meaning of Maury's work. She was finally dismissed.

39 Hertzsprung (1905, 1907).
40 Schwarzschild (1909), p. 449.

Schwarzschild also met Henry Norris Russell (1877–1957) who first learnt of Hertzsprung's publications (1905, 1907). Russell was very impressed, and developed then his diagramme (1912/14). Even here – despite the limited data – the position of the main sequence can be recognised quite well.

Finally, it was called the Hertzsprung-Russell diagramme (HRD), developed by Hertzsprung in 1905/07, and by Russell in 1912/14. Astrophysicists soon recognised its great importance for the discussion of stellar evolution.

Figure 1.16:
Henry Norris Russell (1877–1957) and his first HRD (1914)

(Russell (1914), CC)

1.8 Solar Eclipse Expedition (1905) and the Rise of Solar Physics

For a total solar eclipse on 30 August 1905 with a totality of 3.6 min, Karl Schwarzschild put together an expedition to Algeria, Dixie Camp in Guelma.[41] This expedition was very successful, and he returned with extensive material. During the eclipse, he took photos of the flash spectrum and spectra of the prominences at the edge of the Sun. The flash spectrum contains the emission lines of the chromosphere (lines Ca H and K) at the moment of the complete occultation of the Sun by the Moon. This is the first time that the stratification of the chromosphere has been investigated. It was Schwarzschild's introduction to the subject of solar physics.

In addition, together with Walter Villiger (1872–1938) a Carl Zeiss engineer, Schwarzschild published on Ultraviolet (UV) radiation emitted by the Sun.[42]

Inspired by this, Schwarzschild began his work in the field of theoretical solar physics, introducing the concept of radiative equilibrium for the solar photosphere in 1906:

> *"In granulations, sunspots and prominences, the surface of the sun shows us changing states and stormy changes. In order to understand the physical conditions under which these phenomena occur, it is customary to replace the spatial and temporal changes with an average stationary state, a mechanical equilibrium of the solar atmosphere. (Up to now, the focus has been on the so-called adiabatic equilibrium that prevails in our atmosphere when it is thoroughly mixed by ascending and descending currents).*
> *I would like to draw attention here to another type of equilibrium, which can be described as 'radiative equilibrium'. Radiative equilibrium will occur in a strongly radiating and absorbing atmosphere, in which the mixing effect of ascending and descending currents* [convection] *recedes the heat exchange by radiation* [heat exchange more by radiation than by convection].[43]

41 Schwarzschild: Ueber die totale Sonnenfinsternis (1907), p. 3–73.

42 Schwarzschild & Villiger: (1906), S. 284–305. The Infrared (IR) range was also analysed.

43 *„Die Sonnenoberfläche zeigt uns in Granulationen, Sonnenflecken und Protuberanzen wechselnde Zustände und stürmische Veränderungen. Um die physikalischen Verhältnisse zu begreifen, unter denen diese Erscheinungen stehen, pflegt man in erster Annäherung den räumlichen und zeitlichen Wechsel durch einen mittleren stationären Zustand, ein mechanisches Gleichgewicht der Sonnenatmosphäre zu ersetzen. (Im Vordergrund der Betrachtung stand bisher das sogenannte adiabatische Gleichgewicht, wie es in unserer*

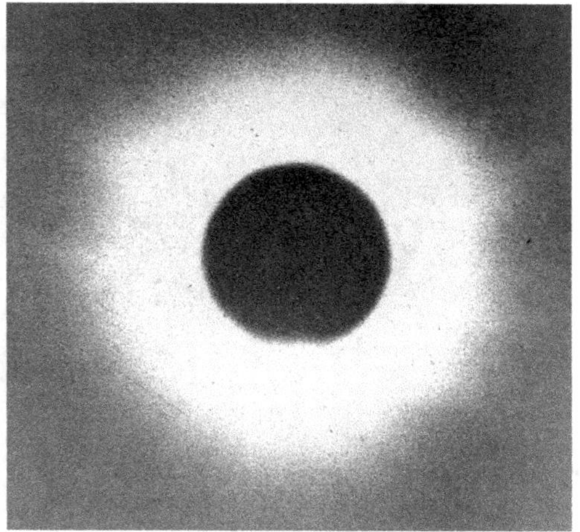
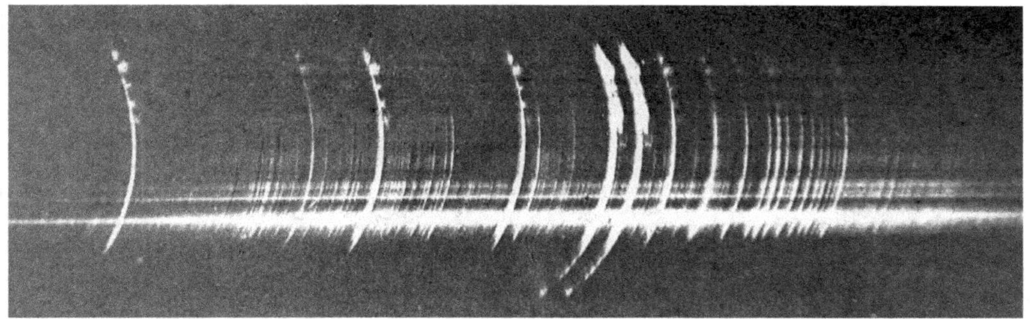

Figure 1.17:
Schwarzschild's total solar eclipse, Guelma/Algeria, 30 August 1905:
Top left: Karl Schwarzschild, servant Ali, and Carl Runge (1856–1927),
Top right: Photograph of the corona of the eclipsed Sun (positive),
Bottom: Flash spectrum of the chromosphere
in the region of the Calcium resonance lines H and K

(Photo: Robert Emden, Schwarzschild (1907c), Taf. I, Fig. 1+2, Taf. III, 17)

Atmosphäre herrscht, wenn sie von auf- und absteigenden Strömungen gründlich durch-

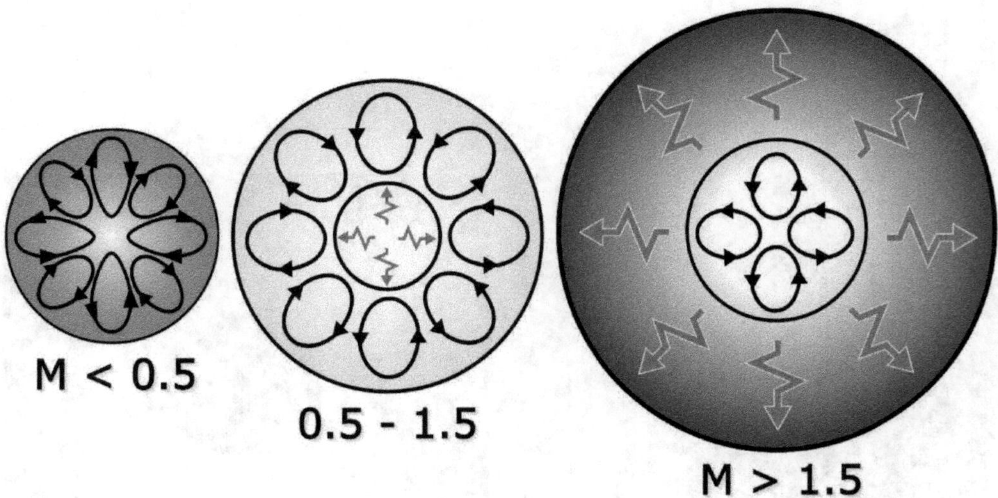

Figure 1.18:
Heat transfer by radiation or convection in the Sun or in cooler and hotter stars – *Schwarzschild criterion*: convective or radiative equilibrium

(CC3, Xenoforme)

The result is that the equilibrium of the solar atmosphere (photosphere) corresponds mainly to the radiation equilibrium (1906). Gravitation and radiation pressure are in equilibrium. The *Schwarzschild criterion* can be used to decide whether convective equilibrium or radiative equilibrium exists. The heat transfer is more by radiation than by convection in the Sun.

1.9 Theory of Stellar Structure

Robert Emden (1862–1940) became interested in star formation and stellar atmospheres as a meteorologist and airship driver. Emden was married to Karl

---

mischt ist.)
Ich möchte hier auf eine andere Art des Gleichgewichts aufmerksam machen, welche man als 'Strahlungsgleichgewicht' bezeichnen kann. Strahlungsgleichgewicht wird sich in einer stark strahlenden und absorbierenden Atmosphäre einstellen, in welcher die durchmischende Wirkung auf- und absteigender Ströme [Konvektion] *gegenüber dem Wärmeaustausch durch Strahlung zurücktritt* [Wärmeaustausch mehr durch Strahlung als durch Konvektion]. Schwarzschild: Gleichgewicht der Sonnenatmosphäre (1906), p. 41–53. Translated by the author.

Figure 1.19:
Radiative equilibrium, drawn in 1906 by Karl Schwarzschild's brother Alfred

(© Iain Hercus (Alfred's grandson), Australia, 2013)

Schwarzschild's sister. This also resulted in close scientific contacts. Emden wanted to apply the mechanical theory of heat to cosmological and meteorological problems.

The theory of radiative equilibrium in the solar atmosphere developed by Schwarzschild was applied by Emden to the Earth's atmosphere, and he succeeded in explaining why the convective troposphere suddenly merges into the stratosphere, which is in radiative equilibrium, from an altitude of around 10 km.[44]

Inspired by Schwarzschild, who had introduced the concept of radiative equilibrium, Emden wrote his groundbreaking book Emden, Robert: *Gaskugeln. Anwendungen der mechanischen Wärmetheorie auf kosmologische und meteorologische Probleme* (1907, Gas Spheres – Applications of Mechanical Heat Theory to Cosmological and Meteorological Problems), in which he was the first to consistently apply the results of thermodynamics to the investigation of the internal structure of the Sun and stars.

Emden saw the Sun and the stars as polytropic *gas spheres* – in equilibrium of internal thermal pressure and gravity. In polytropic spheres, the radiating layers are in radiative equilibrium.

> „*The structure of the Sun from the actual solar ball, in which the mass arrangement is primarily caused by the isentropic* [if the entropy S does not change] *change of state of the masses moving along the radius and the photospheric layers above, which consist of thin, more transparent gases that move in the powerful radiation field of the deeper, hotter layers, and thus come close to radiation equilibrium, offers no fundamental difficulties.*"[45]

The idea of gas spheres was a revolutionary insight at the time, after 100 years in which stars were predominantly regarded by most astronomers, even by Kirchhoff, as solid or liquid bodies; the only exceptions were Jonathan Homer Lane (1819–1880) and August Ritter (1826–1908), who already suggested that stars could be gaseous.

Emden's work formed the basis for Eddington's *Theory of the internal structure of stars* (1926). Arthur Stanley Eddington (1882–1944) started to extend Schwarzschild's earlier work on radiation pressure in Emden polytropic models. Instead of only using the internal thermal pressure, Eddington introduced the radiation pressure which was necessary to prevent the collapse of the gaseous sphere. In addition, Eddington used the radiative equilibrium in comparison

44 Emden: Über Strahlungsgleichgewicht und atmosphärische Strahlung, 1913, p. 55–142.
45 Emden: Gaskugeln 1907, here p. 420.

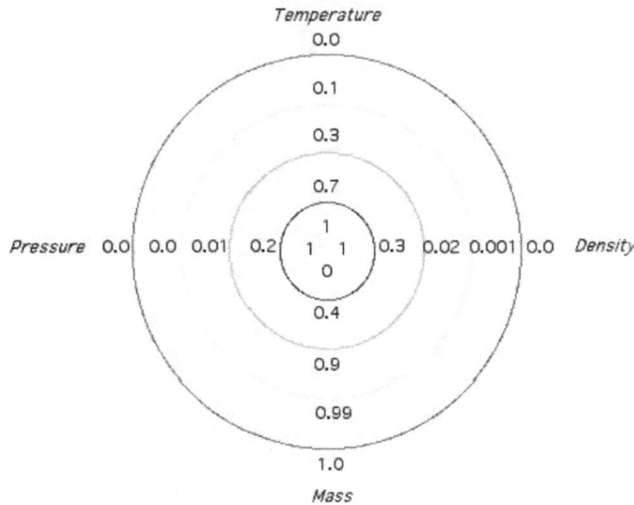

Figure 1.20:
Robert Emden (1862–1940) and Arthur Stanley Eddington (1882–1944)
Standard (twodimensional) Model of Stellar Structure after Emden

(CC, CC)

to Schwarzschild not only in the outer atmospheric layer of the Sun, in the photosphere, but also in the interior of the star.

In 1925, Eddington started to formulate the basic equations of stellar structure,[46] and included his considerations on radiation equilibrium and radiation transport in the model calculations of gas spheres, introduced by Emden. The aim was to describe the layering of stellar matter, i. e. the pressure, temperature, density, as well as luminosity, energy production, and chemical composition with increasing depth.[47]

Eddington speculated already in 1925 about subatomic sources of stellar energy:[48] *"Eddington pointed to the fusion of four hydrogen atoms into a helium atom as the likely energy source supplying the observed stellar luminosity."*[49] The theory of stellar formation, based on Karl Schwarzschild and Eddington, entered a new phase in 1937/38 with the discovery of nuclear fusion as an energy source by Hans Bethe (1906–2015) and Carl Friedrich von Weizsäcker (1912–2007) (CNO-Zyklus, in massive and hot stars). The theory of the internal structure of a star was continued by Subrahmanyan Chandrasekhar (1910–1995), Lyman Spitzer (1914–1997), and Martin Schwarzschild (1912–1997), professor at Princeton University from 1947 to 1979 (*Structure and Evolution of the Stars*, 1958).

Eddington's standard model of a gas sphere (with the polytropes n=3) existed for around thirty years – until computers were available for more precise model calculations from the 1950s onwards.

To summarise, Schwarzschild's work on radiation theory formed an important basis for the theory of stellar structure and evolution as well as for the theory of stellar atmospheres.

1.10 Analysing Stellar Spectra – Theory of Stellar Atmospheres

Kirchhoff's law of thermal radiation was the key to radiation physics and quantum theory,[50] which Max Planck (1858–1947), Kirchhoff's successor as full professor in Berlin, developed in 1900. According to Planck's radiation law, hot objects (stars) are brighter and bluer than cool red ones.

46 Eddington: Internal Constitution of the Stars, 1926. Stars and Atoms, 1927.
47 Schwarz 2011. Cf. Cowling: Development of the theory of stellar structure, 1966.
48 Eddington (1925), p. 419–420.
49 Cf. Mestel (2004), p. 65–73. This is the basic idea for the proton-proton chain for Sun-like stars (or stars with smaller masses).
50 Kangro 1970.

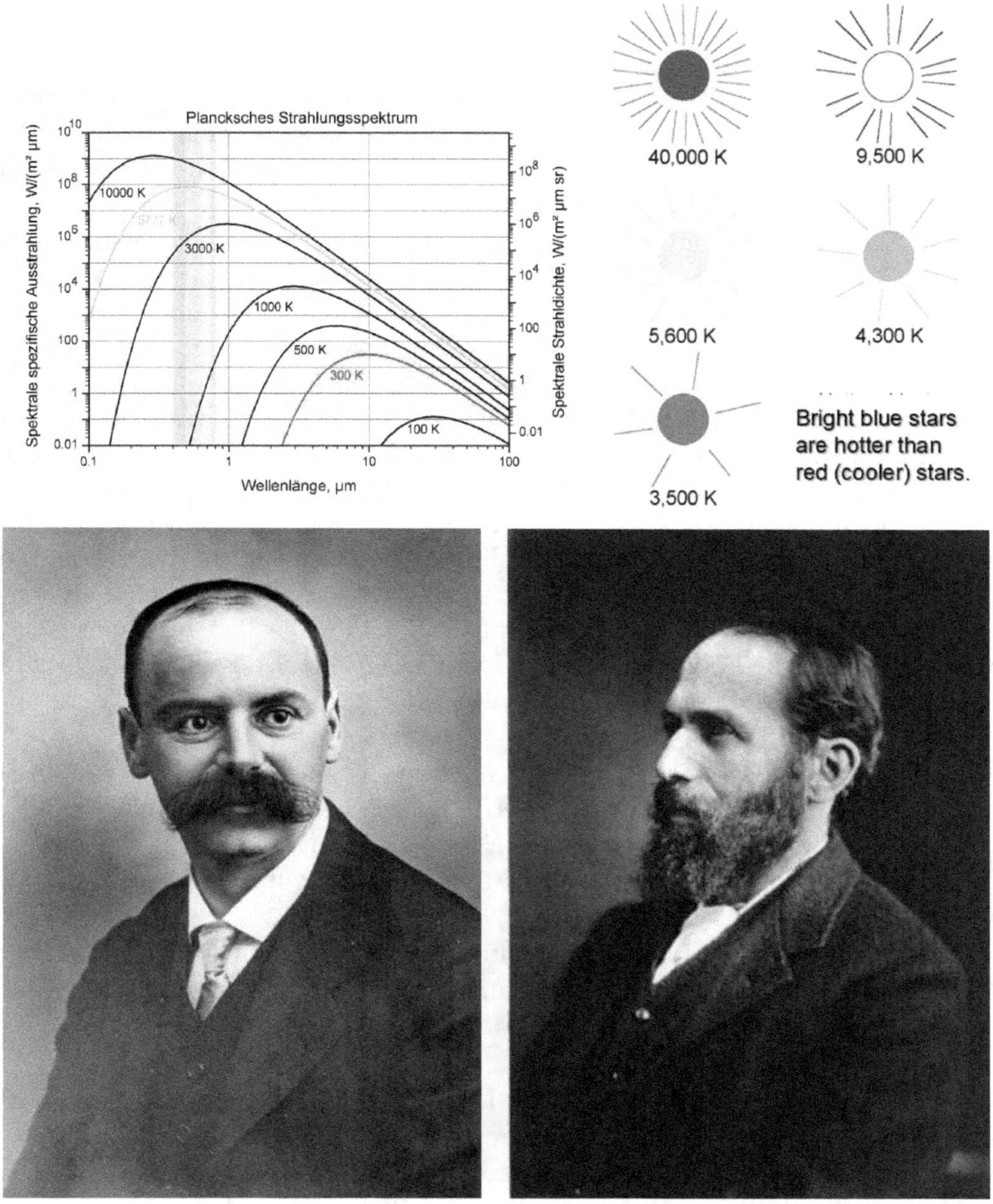

Figure 1.21:
Planck's radiation law (1900), hot stars are brighter and bluer than cool red ones.
Karl Schwarzschild (1873–1916) and Arthur Schuster (1851–1934)

Schwarzschild's introduction of physical concepts into astronomy was expressed in particular in the development of a theory for stellar atmospheres and stellar structure. Key characteristics of stars: local density, local pressure, local surface temperature, local chemical composition. Furthermore, the following are considered: local luminosity, local energy production rate, local opacity, mass within a sphere with radius r.

Karl Schwarzschild and Arthur Schuster (1851–1934) soon applied the spectral energy distribution of a black body's radiation to stellar atmospheres. Schwarzschild succeeded – at least for the photosphere – in establishing a connection between the long observed, but not understood, drop in light from the centre of the solar disc to the edge (centre-edge darkening), and the temperature gradient along the radius – thus building a bridge between theory and observation. The *Schuster-Schwarzschild model* is the basis for analysing stellar spectra in order to calculate the stellar atmosphere. This simple two-layer model was developed by Arthur Schuster in 1905 and Karl Schwarzschild independently of each other in 1906. According to this, at the bottom is the photosphere; there should only be continuous absorption, which emits a continuous spectrum. Above this photosphere lies the so-called *reversing layer*, which produces the Fraunhofer absorption lines. In the Schuster-Schwarzschild model, this layer is based on the assumption of pure scattering and pure absorption.[51]

In the improved *Milne-Eddington model*, there is line scattering and absorption. More general models with scattering, which can be treated analytically, can be solved with the *Schwarzschild-Milne integral equation* in closed form using known functions. We are interested in the intensity of the radiation $J(t, r)$, where t is the optical depth.

The abundance of hydrogen in the interior of the star was not well known at the beginning of the 1930s (due to the uncertainty of the absorption coefficient of the metals). Finally, the analysis can be compared with spectroscopic data on the abundance of various elements.

Cecilia H. Payne-Gaposchkin (1900–1979) focussed her work on theoretical studies and investigated the physical reasons why stars are variable. She received her Ph.D. from Radcliffe College, Harvard (1925): *Stellar Atmospheres. A Contribution to the Observational Study of High Temperature in the Reversing Layers of Stars.*[52] She had to fight for a long time for recognition of her scientific achievements until she finally became the first female professor and director at Harvard, Cambridge, Mass., in 1956.

51 Cf. Unsöld: Physik der Sternatmosphären, 1938.

52 Payne-Gaposchkin: The Stars of High Luminosity, 1930.

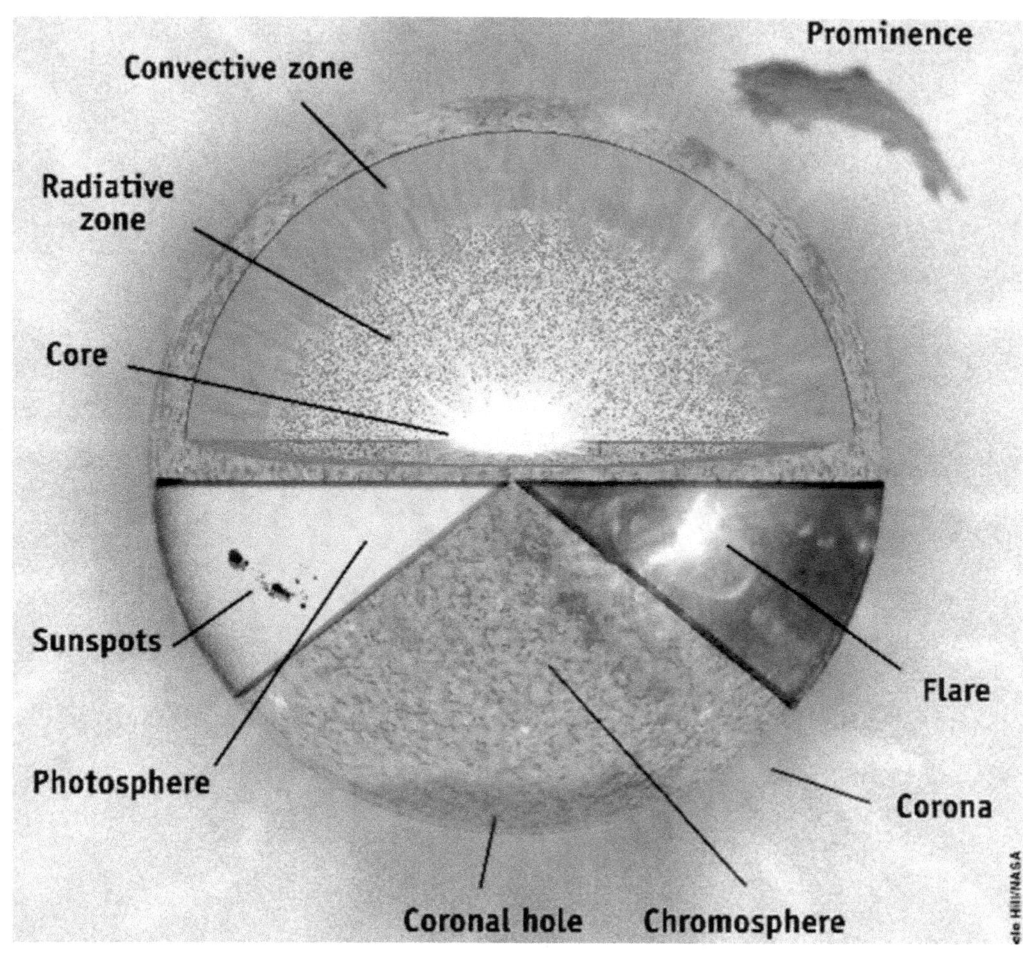

Figure 1.22:
An illustration of the Sun's structure and zones
(credit: ESA & NASA/SOHO)

Cecilia Payne showed that the great variability in the intensity of stellar absorption lines reflects the different proportions of ionisation at different temperatures – and not the elemental proportions. The intensity of absorption lines in stellar atmospheres is a complex function of temperature, pressure, and atomic constants. Based on the ionisation equation of Megh Nad Saha (1894–1956), she determined the chemical abundances in particular from the spectra of the stars.

Figure 1.23:
Cecilia Payne-Gaposchkin (1900–1979)

(© Painting by Patricia Watwood,
Credit: President and Fellows of Harvard College)

Payne lists the abundances of elements on earth in the following order: oxygen, silicon, hydrogen, aluminium, sodium, calcium, iron, magnesium, potassium, titanium, carbon, chlorine, phosphorus, sulphur, nitrogen, manganese, fluorine, chromium, etc. At that time, it was expected that abundances on Earth correspond to those in the Sun or in stars. But she found in her doctoral thesis in 1925 that the Sun consists mainly of hydrogen and helium.

Hydrogen is even the most common element in the cosmos by a factor of about 1 million – an idea that was not recognised by Henry Norris Russell (1877–1957), her doctoral supervisor, until 1929, when he came to this conclusion himself. Otto Struve (1897–1963), Professor at the University of Chicago since 1930, on the other hand, praised Payne's doctoral thesis as *„undoubtedly the most brilliant ever written in astronomy."*

Russell determined the mass ratio in 1929, using a different method to Cecilia Payne,

$$H : He = 3 : 1 \quad \text{(mass ratio)}$$

Henry Norris Russell[53] proposed that only six of the metallic elements (Na, Mg, Si, K, Ca, and Fe) contribute 95% of the whole mass.

Bengt Strömgren (1908–1987) found the relative abundance of hydrogen to be nearly 70%, and helium to be about 27%.[54]

Russell presented in his book Russell et al.: *Astrophysics and Stellar Astronomy* (1926/27) the idea that stellar properties like radius, surface temperature, luminosity, etc., were mainly determined by the star's mass and chemical composition (*Vogt-Russell theorem*).[55] This chemical composition gradually changes with age, with stellar evolution.

1.11 General Theory of Relativity (Tests, Schwarzschild Radius, Black Holes)

Today, Schwarzschild's name is primarily known in connection with *Black Holes.* But already in August 1900, the conference of the *Astronomische Gesellschaft* (Astronomical Society) took place in Heidelberg. At Seeliger's request, Schwarzschild gave a lecture entitled *Ueber das zulässige Krümmungsmaaß des Raumes* (On the measure of curvature of space), thus discussing the possibility of a non-Euclidean space.

From 1911, Albert Einstein (1879–1955) worked intensively on the *General Theory of Relativity*, which he published in November 1915 and 1916.[56] For the

53 Russell: On the Composition of the Sun's Atmosphere (1929), p. 11–82, here p. 54. Today the following values are assumed: Hydrogen (71% of the mass, 91.2% of the atoms) and helium (27.1% of the mass, 8.7% of the atoms). All other elements together make up about 0.1%.

54 Strömgren: The —opacity of stellar matter and the hydrogen content of the stars (1932), p. 118–152.

55 Russell (1931). Heinrich Vogt (1890–1968) discovered the *Vogt-Russell theorem* independently. Vogt (1926).

56 Einstein 1915a, 1915c, 1916a.

English speaking countries, Willem de Sitter (1872–1934) and Arthur Stanley Eddington presented introductions to Einstein's theory.[57]

1.11.1 Einstein's Three Tests for the *General Theory of Relativity*

Einstein proposed three methods for testing the *General Theory of Relativity*:[58]

1. Mercury's Perihelion Rotation (43″ per century)
   Mercury's elliptical orbit around the Sun is not closed, but the perihelion (the point closest to the Sun) orbits the Sun. This had already been discovered by Urbain LeVerrier (1811–1877) in 1859, and can largely be attributed to the Newtonian gravitation of the other planets. However, there remained a small unexplained residue that could only be satisfactorily explained with Einstein's theory.[59] Schwarzschild provided a stringent solution to the perihelion of Mercury (December 1913).[60]
2. Gravitational Redshift (2×10^{-6} of the Wavelength or 0.01°Å)
   In the gravitational field of a celestial body, light should lose energy, which manifests itself in an increase in the wavelength.[61] Especially Karl Schwarzschild[62]

   > *"recognised the fundamental importance of Einstein's theory early on, and did his utmost to promote the plans to prove the relativistic effects. In 1913/14, he himself undertook experiments at the Potsdam Observatory to measure the redshift at the cyan band in the violet part of the solar spectrum – at 0.6 km/s as Einstein had predicted – using a spectrograph system installed on the roof of what was later to become the library building. The results presented by Einstein to the Academy at the end of 1914 [...] made it clear that quantitatively reliable results could only be achieved with much more elaborate means."*[63]

57 Sitter: Space, Time, and Gravitation, (1916), Sitter: On Einstein's Theory of Gravitation and its Astronomical Consequences, (1915–1916), (1916–1917), (1917–1918). Eddington: Space, Time and Gravitation: An Outline of the General Relativity Theory, 1920.

58 Wolfschmidt: Prüfung der Einsteinschen Allgemeinen Relativitätstheorie, 2011, p. 132–169.

59 Einstein 1915b.

60 Voigt: Karl Schwarzschild – Gesammelte Werke, 1992, Vol. 1, p. 36

61 Einstein: Über den Einfluß der Schwerkraft auf die Ausbreitung des Lichtes (1911).

62 Schwarzschild (1914), p. 1183–1200. Schwarzschild (1915a), p. 1201–1213, (1915b).

63 *„erkannte frühzeitig die fundamentale Bedeutung der Einsteinschen Theorie und förderte nach Kräften die Pläne zum Nachweis der relativistischen Effekte. Er selbst unternahm*

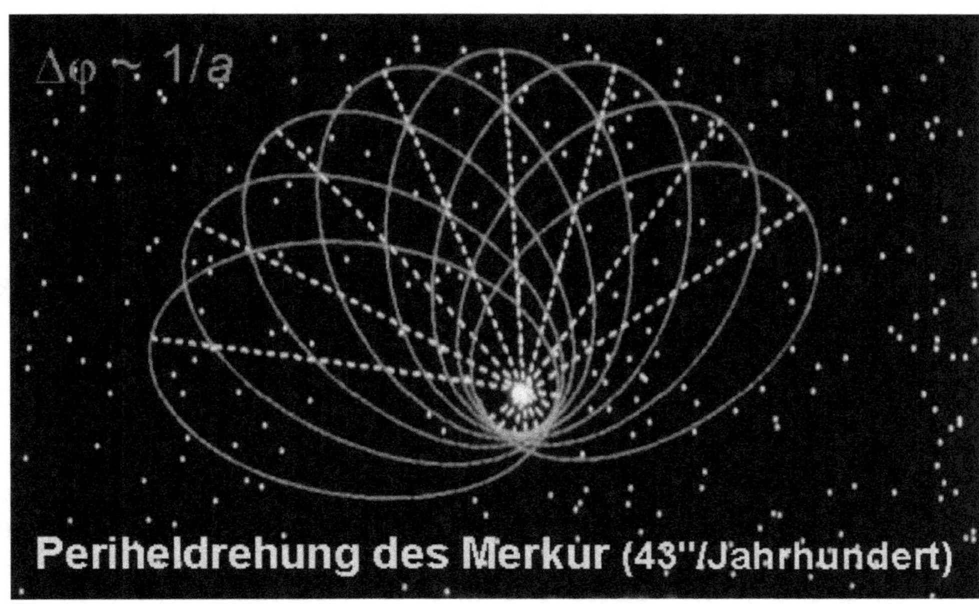

Figure 1.24:
Top: Mercury's Perihelion Rotation
Bottom: Schwarzschild's Test of the Gravitational Redshift
on the Library Building of the Astrophysical Observatory Potsdam

(Photos: Gudrun Wolfschmidt)

Suitable instruments first had to be created to measure the gravitational shift. This was the reason for the construction of the Einstein Tower in Potsdam, enormously promoted by Freundlich's activities (see "3. Deflection of Light").

Eddington recognised *White Dwarfs* like Sirius B, the companion of the very bright star Sirius (α Canis Majoris),[64] as very useful for analysing the gravitational shift because their extraordinary densities: White Dwarfs are *"strange objects that insist on showing a spectrum that in no way corresponds to their luminosity, they can ultimately teach us more than an army of stars that radiate according to the general rule. It seemed that the radiation of the white dwarfs must have been taken for one of those paradoxes which are apt to appear from time to time when one attempts to explain the facts of observation in the absence of sufficient theoretical knowledge."*[65] We know now, that Sirius B has 98% of the mass of the Sun, but only 2.7% of its luminosity, and an extremely high density.[66]

Already in 1924, Eddington calculated and theoretically predicted a gravitational redshift[67] of around +20 km/s on the basis of the temperature (spectral type F0: 8000°) and radius (19,600 km) values assumed at that time, using the *White Dwarf* Sirius B. Then Eddington asked Walter Sidney Adams (1876–1956) of Mt. Wilson Observatory to measure the effect with the 2.5-m-Hooker telescope, the then largest in the world. Finally in

---

1913/14 am Potsdamer Observatorium Versuche zur Messung der Rotverschiebung an der Cyan-Bande im violetten Teil des Sonnenspektrums – um 0,6 km/s wie von Einstein vorhgergesagt – mit einer auf dem Dach des nachmaligen Bibliotheksgebäudes installierten Spektrographenanlage. Die Ende 1914 von Einstein der Akademie vorgelegten Ergebnisse [...] machten deutlich, daß quantitativ zuverlässige Resultate nur mit wesentlich aufwendigeren Mitteln zu erreichen sein würden." Jäger: Der Einstein-Turm in Potsdam und die Relativitätstheorie, (1986), here p. 16.

64 The double star character was suspected in 1834 by Friedrich Wilhelm Bessel (1784–1846) in Königsberg (today Kaliningrad, Russia); in 1944, he noticed irregularities in Sirius' motion by an invisible "dark" companion, disturbing its motion caused by the gravitational attraction. The famous confirmation (1862) was the result of testing a new large lens for a 18½-inch-refractor by Alvan Graham Clark (1832–1897).

65 *„sonderbare Objekte, die sich darauf versteifen, ein Spektrum zu zeigen, das in keiner Weise ihrer Leuchtkraft entspricht, können uns letzten Endes mehr lehren, als ein Heer von Sternen, die nach der allgemeinen Regel ausstrahlen. Es hatte den Anschein, daß man die Strahlung der Weißen Zwerge für eines jener Paradoxien halten mußte, die von Zeit zu Zeit aufzutauchen pflegen, wenn man bei ungenügenden theoretischen Kenntnissen die Beobachtungstatsachen zu erklären versucht."* Eddington 1928, p. 207–208.

66 Eddington continued to puzzle over how a *White Dwarf* can get into a state of extreme compression from which it cannot escape. The star will need energy to cool down.

67 Eddington: Relation between the Masses and Luminosities of the Stars, (1924), p. 308–332.

Figure 1.25:
Sirius A and its faint, tiny stellar companion, Sirius B (bottom left)

(Hubble's Wide Field Planetary Camera – NASA, ESA,
H. Bond (STScI), and M. Barstow, University of Leicester)

1925, Adams (1876–1956) succeeded in measuring a gravitational redshift of about +19 km/s.[68]

Subrahmanyan Chandrasekhar (1910–1995) further developed the theory of *White Dwarfs*. Their behaviour is determined by a degenerate electron gas. In the mid-1930s, he was able to show that above a certain mass, which is 1.4 solar masses, relativistic calculations must be made. In the

68 Adams: The relativity displacement of the spectral lines in the companion of Sirius, (1925), p. 337. Cf. Hetherington: Sirius B and the Gravitational Redshift, (1980), p. 246–252. The result of modern measurements is according to Greenstein et al., (1971) 89 ± 16 km/s.

case of *White Dwarfs* – in contrast to Black Holes – photons can still leave the gravitational field of the star with a loss of energy; the result is the measured gravitational redshift.

3. Deflection of Light in the Gravitational Field of the Sun (1.75″)
John Michell (1724–1793) already suspected an influence of gravity on light in 1784.

Einstein's first prediction of the deflection of light in the gravitational field of a celestial body can be found as early as 1911. From as early as 1913, Erwin Finley Freundlich (1885–1964) had been continuously discussing the *General Theory of Relativity.*[69] In a 1913 application to the Prussian Academy of Sciences (AdW), his early commitment to the astronomical testing of Einstein's General Theory of Relativity also becomes clear:

> „*Following a personal suggestion by Mr Einstein, I took over the examination of this question two years ago* [1911], *and initially attempted to reach a decision on this question with the help of the available material on eclipse photographic plates. At my request, the American and English observatories placed their material at my disposal, and I began to measure the plates with a measuring apparatus from the Astrophysical Observatory in Potsdam, which Prof. Schwarzschild gave me.*"[70]

The measurement during the solar eclipse expedition to the Crimea in 1914 failed due to the outbreak of World War I. Also the measuring of this effect at the Einstein Tower was not successful, because there are too many other influences, i. e. line shift effects.[71]

Arthur Stanley Eddington led a solar eclipse expedition (29 May 1919) to Príncipe Island, West Africa, that provided the first confirmation of

69 Freundlich, who had completed his doctorate under Felix Klein (1849–1925), was a good discussion partner for Einstein concerning problems with the General Theory of Relativity.

70 „*Einer persönlichen Anregung des Herrn Einstein folgend, hatte ich vor zwei Jahren die Prüfung dieser Frage übernommen und vorerst versucht, mit Hilfe des vorhandenen Materials an Finsternisplatten eine Entscheidung dieser Frage zu fällen. Die amerikanischen und englischen Sternwarten stellten mir auf mein Ersuchen ihr Material zur Verfügung, und ich begann mit einem Meßapparat des Astrophysikalischen Observatoriums in Potsdam, den mir Prof. Schwarzschild überließ, die Platten auszumessen.*" Freundlich, Erwin Finlay: Request to the Academy of Sciences dated 7 December 1913. Document no. 88–116. Quoted from: Kirsten & Treder 1979, p. 164–197, here p. 165.

71 In 1913/14, John Evershed (1864–1956), director of Kodaikanal Observatory in India, and Charles Edward St. John (1857–1935) at the Mt. Wilson Observatory also endeavoured in vain to provide clear evidence of the gravitational effect, and to separate it from other causes of the redshift, only partially known at the time. For a detailed discussion of this topic see Hentschel: The Conversion of St. John, (1993), p. 137–194.

Figure 1.26:
Einstein Tower, founded in 1924,
in the Astrophysical Observatory Potsdam-Telegraphenberg

(Free Art License, A. Savin, Wikipedia)

Einstein's *General Theory of Relativity* that gravity bends the path of light when it passes near a massive star. During this total solar eclipse, it was found that the positions of stars, seen just beyond the eclipsed solar

Figure 1.27:
Light deflection at the edge of the Sun during the total solar eclipse 1919: Eddington in Príncipe, West Africa (1919), and Dyson in Sobral, Brazil

(Illustrated London News, Nov. 22, 1919)

disk, were – as the General Theory of Relativity had predicted – slightly displaced away from the centre of the solar disk. The results of the two British expeditions by Eddington (Príncipe, Gulf of Guinea, West Africa), and by Frank Watson Dyson (1868–1939) & Charles Rundle Davidson (1875–1970) (Sobral, Brazil) were presented to the Royal Society and Royal Astronomical Society as a verification of Einstein's *General Theory of Relativity* on 6 November 1919.[72]

Figure 1.28:
Erwin Finlay Freundlich (1885–1964) at the solar eclipse in Sumatra (1929) to measure the relativistic deflection of light;

(Freundlich 1929)

72 Dyson: On the opportunity afforded by the eclipse of 1919 May 29 of verifying Einstein's Theory of Gravitation, (1917). Eddington: The Total Eclipse of 1919, May 29, and the Influence of Gravitation on Light, (1919). Dyson; Eddington & Davidson: Determination of the Deflection of Light by the Sun's Gravitational Field, (1920). Cf. Gilmore & Tausch-Pebody (2022).

1.11.2 First Solution of Einstein's Equation – Schwarzschild Radius

In addition to the attempt to prove the gravitational red shift of spectral lines, Karl Schwarzschild made another important contribution to Einstein's *General Theory of Relativity*:

Another highlight followed: on 13 Januar 1916 – only two months after Einstein's publication of the Field Equations, Schwarzschild provided the first solution. It was a solution for a single spherical, non-rotating (electrically neutral) mass (so-called "outer solution"), it described the gravitational field outside of a spherical mass. This is popularly known as Schwarzschild's *Black Hole.*[73] Schwarzschild had realised that the force of attraction, i. e. the curvature of space-time, can become so strong that not even light can escape. Albert Einstein was extremely impressed by Schwarzschild's achievement; he would never have thought that an exact solution could be formulated so "simply". In this context, the *Schwarzschild Metric* was introduced: in order to describe space-time outside such a sphere, one must develop a metric that tells us that space-time is curved.

The Schwarzschild solution made it possible to specify the radius of a *Black Hole.* Schwarzschild thus introduced the concept of the *Schwarzschild Radius,* the size of the event horizon of a stationary *Black Hole* with *Schwarzschild Radius* using the Schwarzschild coordinates.

$$R_S = \frac{2GM}{c^2}$$

The centre in which the entire mass is united is called the Schwarzschild singularity. A *Black Hole* is a singularity, a super-dense object, from which even light cannot escape. The important property of a *Black Hole* is that the escape velocity has become as great as the speed of light; no particle, not even an electron, no light, no radiation, can leave the *Black Hole,* whose size corresponds to the *Schwarzschild Radius.* Everything that comes close to the Black Hole also crashes into it, causing the escape velocity to increase further, and the *Schwarzschild Radius* to shrink. The radiation undergoes an infinitely large gravitational redshift on the event horizon defined by the *Schwarzschild Radius.*

73 On 24 Februar 1916, Schwarzschild also found a 2nd solution for an expanded, incompressible liquid (so-called "inner solution"). Here Schwarzschild presented the problem for real bodies, he calculated the radius to which an object must be compressed in order to turn it into a Black Hole.

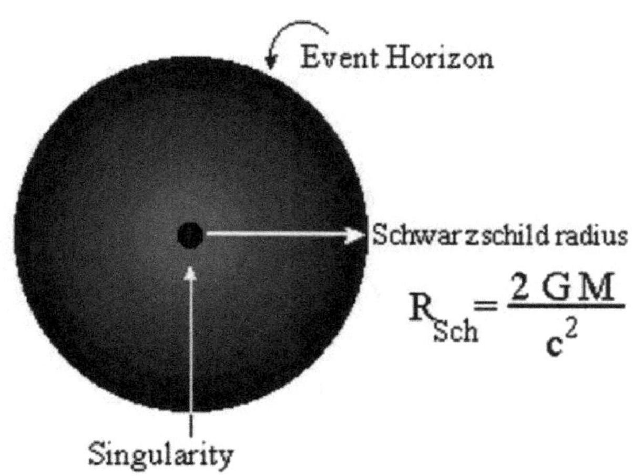

Figure 1.29:
Schwarzschild Radius – Event Horizon of a Black Hole,
Black Hole with accretion disc

(Top: Andrew Schok, University of Alaska Fairbanks, 2019, `http://ffden-2.phys.uaf.edu/webproj/213_fall_2019/Andrew_Schok/andrew_schok/outside.html`, Bottom: NASA)

A *Black Hole* is the final stage of a massive star. Subrahmanyan Chandrasekhar had shown that stars up to an upper limit of 1.46 solar masses end up as *White Dwarfs*; at more than three solar masses, they become *Black Hole.*

- Up to the *Chandrasekhar Limit of 1.46 solar masses*, white dwarfs form.
- Above 1.46 solar masses up to about max. 2.8 to 3 solar masses neutron stars are formed.
- More than 3 solar masses inevitably lead to black holes.

Examples for the size of *Black Holes*:
For the Sun, the Schwarzschild Radius is less than 3 km,
Earth < 1 cm, human 10^{-23} cm,
Black Hole of 3 solar masses $= 6\times10^{30}$ kg – 9 km,
Supermassive Black Hole: 2×10^{36} kg – 3,000,000 km.

However, the concept of a *Black Hole*, the term was first introduced by John Archibald Wheeler (1911–2008) in 1969.

Eddington was also thinking about stars that have more extreme densities than *White Dwarfs*, such as the objects known today as *Black Holes.* This idea goes back to the 18th century, to John Michell (1724–1793),[74] Johann Heinrich Lambert (1728–1777) and Pierre-Simon Laplace (1749–1827). Johann Heinrich Lambert[75] had a dark, supermassive central body as the centre of the Milky Way[76] – the size of Saturn's orbit, with a density greater than that of gold. Lambert introduced this central body by analogy with our solar system in order to ensure the stability of the system; he could probably only imagine a material centre of gravity, but not yet an abstract one:

> *„so I will also have to assume a dark body, which has mass enough to keep the Milky Way in simple order, and if you, my Lord, assume that the Milky Way belongs to countless others, then you give me material for a much heavier and larger body, which must give laws and order to all Milky Ways."*[77]

74 Wolfschmidt 1995. In 1783, John Michell in Thornhill, Yorkshire, after observations with his 30 cm reflector, was of the opinion that nebulae consist of stars or are giant stars. He assumed that the nebulae – which could not be resolved even with the best telescopes – were only very distant systems. Jaki (1977).

75 Lambert: Cosmologische Briefe über die Einrichtung des Weltbaues, 1761; see also Schwarzschild (1907d).

76 The disc-shaped Milky Way contains with millions of stars and has a diameter of 150,000 × the distance of the Sun to Sirius (*Siriusweite*).

77 Lambert 1761, 17th letter; see also 10th letter.

In addition, Pierre Simon de Laplace[78] mentions a body (250 times the diameter of the Sun and with the density of the Earth), whose mass is so great that even light can no longer leave the gravitational field.

1.12 Conclusion and Aftermath – Schwarzschild as Pioneer of Theoretical Astrophysics

Karl Schwarzschild was an outstanding scientist at the turning point from 'classical astronomy' to 'modern theoretical astrophysics'. He is characterized by being very versatile and interdisciplinary – mathematician, physicist, classical astronomer, observational and theoretical astrophysicist.

After starting with some classical astronomy publications on celestial mechanics, positional astronomy, and statistics (Schwarzschild's basic equation of stellar statistics and the *Schwarzschild velocity distribution*), he was active in observational astrophysics in Vienna where he introduced photographic photometry as a quantitative method (*Schwarzschild exponent*). For this contribution, he was honoured with the medal of the *Photographic Society in Vienna* in 1899. In addition, with the colour index, he laid the foundations for multicolour photometry. He was also interested in astronomical instrumentation, and wrote about optics, and proposed a telescope with a large field of view.

Schwarzschild was active in almost all areas of astrophysics, and often published fundamental and groundbreaking work that initiated completely new fields of work. Thus, he is regarded as the founder of theoretical astrophysics.

In the physics of stellar atmospheres, he introduced *radiative equilibrium*. In this field his important contributions are marked by the *Schuster-Schwarzschild model*, the *Schwarzschild-Milne integral equation*, and the *Schwarzschild criterion* for the transition to turbulence. Concerning the rising field of *General Theory of Relativity*, his work in this field can still be seen today in the two solution of the equations, and in the introduction of the *Schwarzschild Radius*.

Hans-Heinrich Voigt Director of Göttingen Observatory from 1963 to 1986, characterised the significance of Schwarzschild as follows:

> *"When an astronomer takes photographic plates, he inevitably has to deal with the Schwarzschild exponent. If he later calculates a stellar atmosphere to analyse his recorded spectrum, he will first start with a Schuster-Schwarzschild model, then perhaps solve the*

78 Laplace: Exposition du système du monde, 1796.

Schwarzschild-Milne integral equation, and finally use the Schwarzschild criterion to see whether he is dealing with convection. The Schwarzschild radius will tell him when the star becomes a black hole. If the astronomer is concerned with the structure of the Milky Way system, he will use the Schwarzschild basic equation of stellar statistics and the Schwarzschild velocity distribution. Even this brief summary clearly shows the great importance of Karl Schwarzschild for astronomy. However, he not only made fundamental contributions to the areas mentioned, but also to the fields of optics, quantum theory and, above all, relativity theory. For example, he provided the first exact solution to the field equations of the general theory of relativity."[79]

Jan Hendrik Oort (1900–1992), the famous Dutch astronomer, described Karl Schwarzschild on the occasion of a memorial ceremony in 1962 as probably the greatest German astronomer since Johannes Kepler (1571–1630).

Einstein particularly emphasised in his memorial speech to Schwarzschild in 1916 that Schwarzschild had created a new astrophysics by introducing many physical subfields such as atomic physics, quantum theory, thermodynamics, radiation theory and relativity theory.

Schwarzschild's admission to the Academy was enormously delayed by the classical astronomer Auwers. It was not until 1912 that he was proposed and elected by Max Planck – with Auwers abstaining. In his programmatic inaugural address at the Berlin Academy of Sciences in 1913, Karl Schwarzschild emphasised the close interaction between physics and astronomy:

„The relationship between astronomy and the other exact sciences is of course particularly close in the field of astrophysics. Doppler's principle gives us the velocities of stars, spectral analysis their chemical composition, and the law of radiation gives us an idea of their temperatures. Astrophysics is sometimes ahead of terrestrial physics. In the nebulae, for example, we know of an element that, judging by its spectral lines, must undoubtedly be an element that astronomers call nebulium, and whose discovery on Earth is the future task of chemists.[80]

79 Voigt, Hans-Heinrich: `http://www.planetarium-goettingen.de/vortragsreihe/3/voigt.html`, accessed 28 December 2024.

80 The greenish line in the spectrum, the *Nebulium* (discovered in 1864), was not a new element. But it was not until 1927 that Ira Sprague Bowen (1898–1973) was able to assign the nebulium line to ionised oxygen (O^{2+}).

Figure 1.30:
Karl Schwarzschild (1873–1916)
Albert Einstein (1879–1955) and Max Planck (1858–1947)

(Top: Göttingen SUB Archive, "Cod. Ms. K. Schwarzschild – 23: 1,2").
(Bottom left: Center for History of Science and Technology (GNT),
University of Hamburg, Bottom right: CC)

> *We currently have a special desideratum for physics. If the mechanism of the glow of the gases could be clarified to such an extent that it could be said with some certainty why the spectral lines of the stars look so extraordinarily different, sometimes narrow, sometimes broad, sometimes sharply defined, sometimes blurred, then we would certainly have made a step forward in understanding the physical conditions in the stellar atmospheres, and perhaps also have gained a more binding judgement about the absolute luminosity of the stars on the basis of their spectral peculiarities. Finding out the absolute luminosity of the stars is such an important goal because knowledge of the absolute luminosity of the stars, combined with their directly observable apparent brightness, results in their distance, and thus a general knowledge of the spatial arrangement of the universe.*
> *Thus there is no end to the lively relationships between astronomy and neighbouring sciences. The clearer this is for the past, the more – you will forgive me for trying to substantiate it – with hopes for the future. Mathematics, physics, chemistry, and astronomy are marching in one front. Those who lag behind are pulled behind. Those who run ahead pull the others. There is the closest solidarity between astronomy and the entire circle of exact sciences."*[81]

Max Planck responded to Schwarzschild's inaugural address in 1913:

> *„Your conception of the task you were given when you assumed your office* [as Director of the Astrophysical Observatory] *is* [...] *permeated by the conviction that neither astronomy nor astrophysics can develop in a sufficiently versatile way unless its relations to general physics and chemistry are cultivated and promoted in such a way that it grows together with them to form a unique whole whose parts complement and support each other. This view has determined the direction of your work from the very beginning of your scientific activity. And the fruits have not failed to materialise.*
> *Whether you investigate the thermodynamic equilibrium of the solar body, the path of light rays from their emission, their pressure on the cosmic mass particles, their passage through the telescope, their blackening effect on the film, you always place physico-chemical research at the service of astronomical research, and thereby reap a rich harvest."*[82]

81 Schwarzschild: Antrittsrede (1913), here p. 598–599.
82 Planck: Erwiderung (1913), p. 600–602.

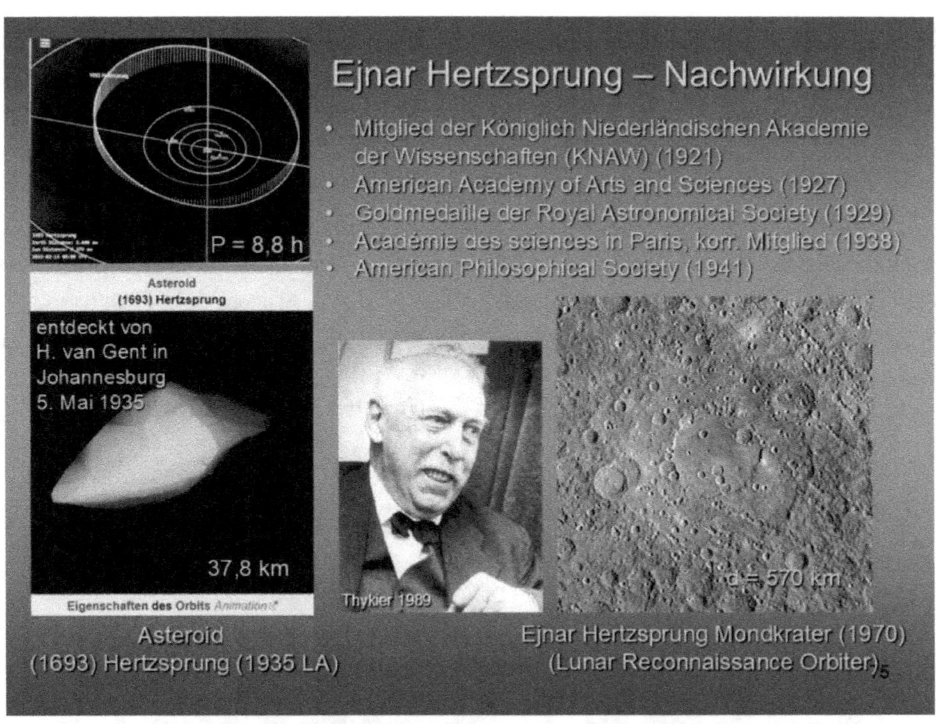

Figure 1.31:
Ejnar Hertzsprung (1873–1967) & Karl Schwarzschild (1873–1916)

Schwarzschild marked the end of the successful era of the genesis of astrophysics, which began with Kirchhoff and Bunsen, continued with Hermann Carl Vogel, and reached its climax with Schwarzschild. There was no suitable successor in the whole of Germany, which Einstein also had to realise with regret.[83] This was the first step towards the decline of astrophysics in Germany, which was lamented in the 1920s.

Both scientists, Hertzsprung and Schwarzschild, received many honours, were accepted into scientific societies, and their names appear on the moon and asteroids. Finally, the world's largest Schmidt telescope in the world with a mirror diameter of 2 m, a free aperture of 1.34 m, and a focal length of 4 m is located at the Karl Schwarzschild Observatory in Tautenburg near Jena.

1.13 Archive Material and Bibliography

1.13.1 Archive Material

BAMBERG, CARL: Katalog Astro 40, Berlin 1924, p. 41, Fig. 27. Deutsches Museum, Sondersammlung, Firmenschriften 271 Askania.

Cod. Ms. K. Schwarzschild (9.10.1873–11.5.1916), Inventarnummern: Acc. Mss. 1965.12. – Acc. Mss. 1979.9. – Acc. Mss. 1990.26. – Acc. Mss. 1990.36. – Acc. Mss. 1992.1. – Acc. Mss. 1992.23. Nachlass K. Schwarzschild; Niedersächsische Staats- und Universitätsbibliothek (SUB) Göttingen. Göttingen 2014.

MÜLLER, GUSTAV: Brief an Karl Schwarzschild vom 18.4.1910 aus Teneriffa. Niedersächsische Staats- und Universitätsbibliothek Göttingen, Handschriftenabteilung, Nachlaß Karl Schwarzschild, Briefe 535, G. Müller

SCHWARZSCHILD, KARL; Voigt, Woldemar; Wiechert, Emil; Wagner, Hermann; Klein, Felix; Riecke, Eduard: Betrifft Einrichtung einer südlichen Filiale der Göttinger physikalisch-astronomischen Institute. 7. August 1905. An den Minister der geistlichen, Unterrichts- und Medicinalangelegenheiten. Universitäts-Sternwarte Göttingen.

SCHWARZSCHILD, KARL: Betrifft Einrichtung eines Observatoriums der kgl. Gesellschaft der Wissenschaften zu Windhuk. 27 July 1908. University Observatory Göttingen.

SCHWARZSCHILD, KARL: Brief an den Minister der geistlichen und Unterrichts-Angelegenheiten vom 7. Juli 1913. Schriftwechsel mit Schmidt in Mittweida, 1910–1916. Archiv der Akademie der Wissenschaften in Berlin, Astro Obs 31.

SCHWARZSCHILD, MARTIN, interviewed by Spencer Weart, Princeton University, March 10, 1977: AIP – `https://www.aip.org/history-programs/niels-bohr-library/oral-histories/4870-1`, accessed December 23, 2024.

83 Einstein: Gedächtnisrede (1916b).

Steinheil, Rudolf: Brief an Karl Schwarzschild vom 1. Feb. und 7. März 1912. Schriftwechsel mit Schmidt in Mittweida, 1910–1916. Archiv der Akademie der Wissenschaften in Berlin, Astro Obs 31.

1.13.2 Bibliography

Adams, Walter Sydney: The Relativity Displacement of the Spectral Lines in the Companion of Sirius. In: *Proceedings of the National Academy of Sciences of the United States of America* **11** (1925), Nr. 7, S. 382–387.

Angenheister, Gustav Heinrich: Geschichte des Samoa-Observatoriums von 1902 bis 1921. In: Birett, H.; Helbig, K.; Kertz, W. & U. Schmucker (ed.): *Zur Geschichte der Geophysik. Festschrift zur 50jährigen Wiederkehr der Gründung der Deutschen Geophysikalischen Gesellschaft.* Berlin: Springer 1974, p. 43–66.

Bruggencate, Paul ten: Karl Schwarzschild – Der Schöpfer der heutigen Astrophysik. In: Schwerte, H. & W. Spengler (ed.): *Forscher und Wissenschaftler im heutigen Europa. Band 1: Weltall und Erde.* München: Oldenburg 1955, p. 232–239.

Cowling, Thomas G.: Development of the theory of stellar structure. In: *Quarterly Journal of the Royal Astronomical Society* **VII** (1966), p. 121–137.

De Vorkin, David H.: Stellar Evolution and the Origin of the Hertzsprung-Russell Diagram. In: *Astrophysics and Twentieth-Century Astronomy to 1950: Part A.* Edited by Owen Gingerich. Cambridge: Cambridge University Press (The General History of Astronomy, Vol. 4A) 1984, p. 90–108.

Dick, Wolfgang R. & Klaus Fritze (Hg.): *300 Jahre Astronomie in Berlin und Potsdam: eine Sammlung von Aufsätzen aus Anlass des Gründungsjubiläums der Berliner Sternwarte.* Thun, Frankfurt am Main: Verlag Harri Deutsch 2000.

Dyson, Frank W.: On the opportunity afforded by the eclipse of 1919 May 29 of verifying Einstein's Theory of Gravitation. In: *Monthly Notices of the Royal Astronomical Society* **77** (1917), p. 445–447.

Dyson, Frank W.: Eddington, Arthur Stanley & Charles Rundle Davidson: Determination of the Deflection of Light by the Sun's Gravitational Field, from Observations Made at the Total Eclipse of May 29, 1919. In: *Philosophical Transactions of the Royal Society. A: Mathematical, Physical and Engineering Sciences* **220** (1920), p. 571–581.

Eddington, Arthur Stanley: The Total Eclipse of 1919, May 29, and the Influence of Gravitation on Light. In: *Observatory* **42** (1919), p. 119–122; p. 389.

Eddington, Arthur Stanley: *Space, Time and Gravitation: An Outline of the General Relativity Theory.* Cambridge: Cambridge University Press 1920.

Eddington, Arthur Stanley: Relation between the Masses and Luminosities of the Stars. In: *Monthly Notices of the Royal Astronomical Society* **84** (1924), p. 308–332.

EDDINGTON, ARTHUR STANLEY: The Source of Stellar Energy. In: *Nature* **115** (1925), p. 419-420.

EDDINGTON, ARTHUR STANLEY: *The Internal Constitution of Stars.* Cambridge, UK: Cambridge University Press 1926.

EINSTEIN, ALBERT: Über den Einfluß der Schwerkraft auf die Ausbreitung des Lichtes. In: *Annalen der Physik* **340** (1911), Nr. 10, p. 898–908.

EINSTEIN, ALBERT: Zur allgemeinen Relativitätstheorie. In: *Sitzungsberichte der Königlich Preußischen Akademie der Wissenschaften* (1915a), p. 778–786, 799–801.

EINSTEIN, ALBERT: Erklärung der Perihelbewegung des Merkur aus der allgemeinen Relativitätstheorie. In: *Sitzungsberichte der Königlich Preußischen Akademie der Wissenschaften* (1915b), p. 831–839.

EINSTEIN, ALBERT: Die Feldgleichungen der Gravitation. In: *Sitzungsberichte der Königlich Preußischen Akademie der Wissenschaften* (1915c), p. 844–847.

EINSTEIN, ALBERT: Die Grundlage der allgemeinen Relativitätstheorie. In: *Annalen der Physik* **49** (1916a), p. 769–822.

EINSTEIN, ALBERT: Gedächtnisrede des Hrn. Einstein auf Schwarzschild. In: *Sitzungsberichte der Königlich Preußischen Akademie der Wissenschaften* (1916b), p. 768–770.

ELLIOTT, IAN: Monck, William Henry Stanley. In: HOCKEY, T., ET AL.: *The Biographical Encyclopedia of Astronomers.* New York, NY: Springer 2007.

EMDEN, ROBERT: *Gaskugeln. Anwendungen der mechanischen Wärmetheorie auf kosmologische und meteorologische Probleme.* Leipzig, Berlin: B. G. Teubner 1907.

EMDEN, ROBERT: Über Strahlungsgleichgewicht und atmosphärische Strahlung. In: *Sitzungsberichte der Mathematisch-Physikalischen Klasse der Bayerischen Akademie der Wissenschaften*, 1913, p. 55–142.

GILMORE, GERARD & GUDRUN TAUSCH-PEBODY: The 1919 eclipse results that verified general relativity and their later detractors: a story retold. In: *Notes and Records* **76** (2022), p. 155–180.

GREENSTEIN, J. L.; OKE, J. B. & H. L. SHIPMAN: Effective Temperature, Radius, and Gravitational Redshift of Sirius B. In: *Astrophysical Journal* **169** (1971), p. 563–566.

HELDEN, ALBERT VAN: Ejnar Hertzsprung, 1873–1967. In: BERKEL, KLAAS VAN; HELDEN, ALBERT VAN & LODEWIJK PALM (ed.): *A History of Science in The Netherlands. Survey, Themes and Reference.* Leiden, Boston, Köln: Brill 1999, p. 460–463.

HENTSCHEL, KLAUS: The Conversion of St. John: A Case Study on the Interplay of Theory and Experiment. In: *Science in Context* **6** (1993), 1, p. 137–194.

HERRMANN, DIETER B.: Julius Scheiner und der erste Lehrstuhl der Astrophysik an der Universität Berlin. In: *NTM – Schriftenreihe für Geschichte der Naturwissenschaft, Technik und Medizin* **14** (1977), p. 33–42.

HERRMANN, DIETER B.: *Ejnar Hertzsprung – Pionier der Sternforschung.* Berlin, Heidelberg, New York: Springer 1994, 2011.

HERTZSPRUNG, EJNAR: Zur Strahlung der Sterne. In: *Zeitschrift für Wissenschaftliche Photographie, Photophysik und Photochemie* **3** (1905), Heft 11, p. 429–442; **5** (1907), p. 86–107.

HERTZSPRUNG, EJNAR: Eine Annäherungsformel für die Abhängigkeit zwischen Beleuchtungshelligkeit und Unterschiedempfindsamkeit des Auges. In: *Zeitschrift für Wissenschaftliche Photographie, Photophysik und Photochemie* **5** (1907), Heft 12, p. 468–472.

HERTZSPRUNG, EJNAR: Über die räumliche Verteilung der Veränderlichen vom δ-Cephei-Typus (On the spatial distribution of variable [stars] of the δ-Cephei-Type). In: *Astronomische Nachrichten* **196** (1913), No. 4692, p. 201–208.

HETHERINGTON, NORRISS S.: Sirius B and the Gravitational Redshift. An Historical Review. In: *Quarterly Journal of the Royal Astronomical Society* **21** (1980), p. 246–252.

HOFFLEIT, E. DORRIT: Pioneering Women in the Spectral Classification of Stars. In: *Physics in Perspective* **4** (2002), p. 370–398.

JÄGER, FRIEDRICH WILHELM: Der Einstein-Turm in Potsdam und die Relativitätstheorie. In: *Die Sterne* **62** (1986), S. 14–23.

JAKI, STANLEY L.: Dunkle Regenten als Vorläufer Schwarzer Löcher. Lambert-Sonderheft aus Anlaß des 200sten Todestages von J. H. Lambert. In: *Nachrichten der Olbers-Gesellschaft Bremen* (1977), Nr. 107, S. 1–10.

KANGRO, HANS: *Vorgeschichte des Planckschen Strahlungsgesetzes, Messungen und Theorien der spektralen Energieverteilung bis zur Begründung der Quantenhypothese.* Wiesbaden: Franz Steiner (Boethius; Bd. 11) 1970.

KIRSTEN, CHRISTA & HANS-JÜRGEN TREDER (ed.): *Albert Einstein in Berlin 1913–1933, Teil I: Darstellung und Dokumente.* Berlin: de Gruyter (Studien zur Geschichte der Akademie der Wissenschaften der DDR; 6)1979.

LAMBERT, JOHANN HEINRICH: *Cosmologische Briefe über die Einrichtung des Weltbaues.* Augsburg: Eberhard Kletts Wittiv 1761.

LAPLACE, PIERRE SIMON DE: *Exposition du système du monde.* Paris: Impremerie du Cercle-Social 1796.

MESTEL, LEON: Arthur Stanley Eddington: Pioneer of stellar structure theory. In: *Journal of Astronomical History and Heritage* **7** (2004), No. 2, p. 65–73.

MEYER, MAX WILHELM: *Das Weltgebäude. Eine gemeinverständliche Himmelskunde.* Leipzig: Bibliographisches Institut 1898.

MONCK, WILLIAM HENRY STANLEY: The Spectra and Colours of Stars. In: *Journal of the British Astronomical Association* **5** (1895), p. 416–419.

PAYNE-GAPOSCHKIN, CECILIA: *The Stars of High Luminosity.* New York, London: McGraw-Hill 1930.

PLANCK, MAX: Erwiderung des Sekretars Hrn. Planck. In: *Sitzungsberichte der Königlich Preußischen Akademie der Wissenschaften* (1913), Nr. 32, S. 600–602.

REINSCH, KLAUS & AXEL D. WITTMANN (Hg.): *Karl Schwarzschild (1873–1916).* Göttingen: Universitätsverlag 2017.

RUSSELL, HENRY NORRIS: Relations Between the Spectra and Other Characteristics of the Stars. In: *Popular Astronomy* **22** (1914), p. 275–294.

RUSSELL, HENRY NORRIS: On the Composition of the Sun's Atmosphere. In: *Astrophysical Journal* **70** (1929), p. 11–82

RUSSELL, HENRY NORRIS: Notes on the Constitution of the Stars. In: *Monthly Notices of the Royal Astronomical Society* **91** (1931), p. 951–966.

RUSSELL, HENRY NORRIS; DUGAN, RAYMOND SMITH & JOHN QUINCY STEWART: *Astronomy: A Revision of Young's Manual of Astronomy. Vol. 1: The Solar System. Vol. 2: Astrophysics and Stellar Astronomy.* Boston, New York: Ginn & Co. 1926/27, 1938, 1945.

SCHEINER, JULIUS: *Die Spectralanalyse der Gestirne.* Leipzig 1890.

SCHLESINGER, FRANK: Correspondence Concerning the Classification of Stellar Spectra. In: *Astrophysical Journal* **33** (1911), p. 260–300.

SCHWARZ, OLIVER: Die frühe Entwicklung der Theorie des inneren Aufbaus der Sterne. In: WOLFSCHMIDT, GUDRUN (ed.): *Entwicklung der Theoretischen Astrophysik.* Hamburg: tredition (Nuncius Hamburgensis; Band 4) 2011, p. 266–281.

SCHWARZSCHILD, KARL: Methode zur Bahnbestimmung der Doppelsterne. In: *Astronomische Nachrichten* **124** (1890), Issue 13, Nr. 2965, p. 215–218.

SCHWARZSCHILD, KARL: Methode zur Bahnbestimmung der Doppelsterne. In: *Astronomische Nachrichten* **124** (1890), Issue 13, Nr. 2965, p. 215–218.

SCHWARZSCHILD, KARL: Beobachtungen von veränderlichen Sternen und der Nova Aurigae. In: *Astronomische Nachrichten* **129** (1892), pp. 399–404.

SCHWARZSCHILD, KARL: Beiträge zur photographischen Photometrie der Gestirne. In: *Publicationen der von Kuffner'schen Sternwarte Wien* **V** (1900a), C3–C135.

SCHWARZSCHILD, KARL: Ein Verfahren der Bahnbestimmung bei spektroskopischen Doppelsternen. In: *Astronomische Nachrichten* **152** (1900b), No. 3629, p. 65–74.

SCHWARZSCHILD, KARL: Die Bestimmung der Sternhelligkeiten aus extrafokalen photographischen Aufnahmen. In: *Publikationen der von Kuffnerschen Sternwarte in Wien* 5 B und 5 C (1900c).

SCHWARZSCHILD, KARL & BRUNO MEYERMANN: Über eine Schraffierkassette zur Aktinometrie der Sterne. In: *Astronomische Nachrichten* **170** (1906), p. 277–282.

SCHWARZSCHILD, KARL & WALTER VILLIGER: On the Distribution of Brightness of the Ultra-Violet Light on the Sun's Disk. In: *Astrophysical Journal* **23** (1906), p. 284–305.

Schwarzschild, Karl: Über das Gleichgewicht der Sonnenatmosphäre. In: *Nachrichten der Königlichen Gesellschaft der Wissenschaften zu Göttingen*, math.-phys. Klasse (1906), Heft 1, p. 41–53.

Schwarzschild, Karl: Über die Eigenbewegung der Fixsterne. In: *Nachrichten der Königlichen Gesellschaft der Wissenschaften zu Göttingen*, math.-phys. Klasse (1907a), Issue 5, p. 614.

Schwarzschild, Karl: Über eine neue Schraffierkassette. In: *Astronomische Nachrichten* **174** (1907b), p. 137–140.

Schwarzschild, Karl: Ueber die totale Sonnenfinsternis vom 30. August 1905. In: *Abhandlungen der Kgl. Gesellschaft der Wissenschaften zu Göttingen* NF 5/2 (1907c), S. 3–73 = *Astronomische Mitteilungen der Kgl. Sternwarte zu Göttingen* **13** (1906), p. 3–73.

Schwarzschild, Karl: Über Lamberts Kosmologische Briefe. In: *Nachrichten von der Königlichen Gesellschaft der Wissenschaften zu Göttingen* (phil.-hist. Klasse) 59 (1907d).

Schwarzschild, Karl: *Betrifft Einrichtung eines Observatoriums der Kgl. Gesellschaft der Wissenschaften zu Windhuk, 27. Juli 1908.* Universitäts-Sternwarte Göttingen 1908.

Schwarzschild, Karl: Über das System der Fixsterne. In: *Himmel und Erde* 21 (1909), S. 433–451.

Schwarzschild, Karl (ed.): *Aktinometrie der Sterne der BD in der Zone 0° bis +20° Deklination. Teil A.* Unter Mitwirkung von Bruno Meyermann, Arnold Kohlschütter & O. Birck. Abhandlungen der Königlichen Gesellschaft der Wissenschaften zu Göttingen, math.-phys. Klasse, Neue Folge VI (1910a), Nr. 6, p. 1–115 und Neue Folge VIII, Nr. 1.

Schwarzschild, Karl: Die großen Sternwarten der Vereinigten Staaten. In: *Internationale Wochenschrift für Wissenschaft, Kunst und Technik* (3. Dez. 1910b), p. 1531–1544.

Schwarzschild, Karl: Ein Verfahren der Bahnbestimmung bei spectroskopischen Doppelsternen [β Aurigae]. In: *Astronomische Nachrichten* **152** (1900c), Nr. 3629, p. 65–74.

Schwarzschild, Karl (ed.): *Aktinometrie der Sterne der BD in der Zone 0° bis +20° Deklination. Teil B.* Unter Mitwirkung von Bruno Meyermann, Arnold Kohlschütter, O. Birck, & Władysław Dziewulski. Abhandlungen der Königlichen Gesellschaft der Wissenschaft zu Göttingen, math.-phys. Klasse, Neue Folge VIII, Nr. 4. Berlin: Weidmannsche Buchhandlung 1912a.

Schwarzschild, Karl: Nova 18.1912 Geminorum (Remark on the spectrum of Enebo's Nova). In: *Astronomische Nachrichten* **191** (1912b), p. 45–50.

Schwarzschild, Karl: Antrittsrede [Inaugural address] des Hrn. Schwarzschild. In: *Sitzungsberichte der Königlich Preußischen Akademie der Wissenschaften* (1913), Nr. 32, S. 596–600.

SCHWARZSCHILD, KARL: Über Diffusion und Absorption in der Sonnenatmosphäre. In: *Sitzungsberichte der Akademie der Wissenschaften zu Berlin*, math.-phys. Kl. (1914) II, (November 1914), p. 1183–1200.

SCHWARZSCHILD, KARL: Über die Verschiebung der Bande bei 3883 Å im Sonnenspektrum. In: *Sitzungsberichte der Akademie der Wissenschaften zu Berlin*, math.-phys. Kl. (1914) II, (Januar 1915a), S. 1201–1213.

SCHWARZSCHILD, KARL: On the displacement of the bands at 3883 Å in the solar spectrum. In: *Proceedings of the Academy of Sciences in Berlin*, math.-phys. Kl. (1914) II, (January 1915b), p. 1201–1213.

SCHWARZSCHILD, KARL: Über das Gravitationsfeld eines Massenpunktes nach der Einsteinschen Theorie. In: *Sitzungsberichte der Königlich Preußischen Akademie der Wissenschaften* (1915c), p. 189–196.

SCHWARZSCHILD, KARL: Über das Gravitationsfeld einer Kugel aus inkompressibler Flüssigkeit nach der Einsteinschen Theorie. In: *Sitzungsberichte der Königlich Preußischen Akademie der Wissenschaften* (1916), p. 424–434.

SCHWARZSCHILD, MARTIN: *Structure and Evolution of the Stars.* Princeton: Princeton University Press 1958.

SITTER, WILLEM DE: Space, Time, and Gravitation. In: *Observatory* **39** (1916), p. 412–419.

SITTER, WILLEM DE: Sitter, Willem de: On Einstein's Theory of Gravitation and its Astronomical Consequences. In: *Monthly Notices of the Royal Astronomical Society* **76** (1915–1916), p. 699–728, **77** (1916–1917), p. 155–184, **78** (1917–1918), p. 3–28.

STRÖMGREN, BENGT: The opacity of stellar matter and the hydrogen content of the stars. In: *Zeitschrift für Astrophysik* **4** (1932), p. 118–152.

TREDER, HANS-JÜRGEN: Karl Schwarzschild und die Wechselbeziehungen zwischen Astronomie und. Physik. In: *Die Sterne* **50** (1974), 50, p. 13–19.

UNSÖLD, ALBRECHT: *Physik der Sternatmosphären, mit besonderer Berücksichtigung der Sonne.* Berlin: Springer 1938; (2nd ed.) 1955.

VOGT, HEINRICH: Die Beziehung zwischen den Massen und den absoluten Leuchtkräften der Sterne. In: *Astronomische Nachrichten* **226** (1926), S. 301.

VOIGT, HANS-HEINRICH: Von Karl Schwarzschild bis Hans Kienle. Der Weg von der klassischen Astronomie zur Astrophysik in Göttingen. In: *Sterne und Weltraum* **28** (1989), p. 12–17.

VOIGT, HANS-HEINRICH (ed.): *Karl Schwarzschild – Gesammelte Werke, Collected Works, Vol. 1–3.* Heidelberg: Springer 1992.

VOIGT, HANS-HEINRICH: Karl Schwarzschild (1873–1916). In: WOLFSCHMIDT, GUDRUN (ed.): *Entwicklung der Theoretischen Astrophysik.* Hamburg: tredition (Nuncius Hamburgensis; Band 4) 2011, p. 244–265.

WILSON, RAYMOND NEIL: Karl Schwarzschild and Telescope Optics. In: *Reviews in Modern Astronomy* **7** (1994), S. 1–29.

Wolfschmidt, Gudrun: *Milchstraße, Nebel, Galaxien – Strukturen im Kosmos von Herschel bis Hubble.* München: Deutsches Museum (Abhandlungen und Berichte, Neue Folge, Band 11), München: Oldenbourg-Verlag 1995.

Wolfschmidt, Gudrun: Die Entwicklung der astronomischen Photometrie bis in die 20er Jahre. In: *Wissenschaftliches Jahrbuch des Deutschen Museums.* München: R. Oldenbourg (Abhandlungen und Berichte N.F. 6) 1989, p. 227–268, p. 271–272.

Wolfschmidt, Gudrun: Nikolaus von Konkoly (1842–1916) als Begründer des Konkoly Observatoriums Budapest. In: Wolfschmidt, Gudrun (ed.): *Astronomisches Mäzenatentum. Astronomical Patronage.* Norderstedt: Books on Demand (Nuncius Hamburgensis; Vol. 11) 2008, p. 82–109.

Wolfschmidt, Gudrun: Gothard and Konkoly – national and international network of science. In: *Conference on the 150th anniversary of Gothard's birth, Gothard Observatory, Szombathely, Hungary, May 31 to June 2, 2007.* Ed. by István Jankovics et al.. Szombathely 2009.

Wolfschmidt, Gudrun (ed.): *Cultural Heritage of Astronomical Observatories – From Classical Astronomy to Modern Astrophysics.* Proceedings of International ICOMOS Symposium in Hamburg, October 14–17, 2008. Berlin: hendrik Bäßler-Verlag (International Council on Monuments and Sites – Monuments and Sites XVIII) 2009.

Wolfschmidt, Gudrun: Prüfung der Einsteinschen Allgemeinen Relativitätstheorie. In: Wolfschmidt, Gudrun (ed.): *Entwicklung der Theoretischen Astrophysik.* Hamburg: tredition (Nuncius Hamburgensis; Band 4) 2011, p. 132–169.

Wolfschmidt, Gudrun: Der Tübinger Astrophysiker Hans Rosenberg und seine photometrischen Arbeiten. In: Wolfschmidt, Gudrun (ed.): *Der Himmel über Tübingen – Barocksternwarten – Landesvermessung – Astrophysik.* Hamburg: tredition (Nuncius Hamburgensis; Band 28) 2014, p. 280–311.

Wolfschmidt, Gudrun: Karl Schwarzschild, der Begründer der theoretischen Astrophysik. In: *Physik in unserer Zeit – PhiuZ* **47** (2016), Heft 6, S. 294–300.

Wolfschmidt, Gudrun: Karl Schwarzschild, bedeutendster Nachfolger von Gauß und Begründer der theoretischen Astrophysik. In: *Mitteilungen der Gauss Gesellschaft e. V. Göttingen* (GGM) **54** (2017), S. 9–27.

Wolfschmidt, Gudrun: Kosmochemie – Chemische Elemente im Kosmos – Meteoriten, Sterne, Kosmologie. In: Wolfschmidt, Gudrun (ed.): *Kosmochemie – Geschichte der Entdeckung und Erforschung der chemischen Elemente im Kosmos. Cosmochemistry – History of Discovery and Research of Chemical Elements in the Cosmos.* Hamburg: tredition (Nuncius Hamburgensis; Vol. 50) 2022, p. 66–135.

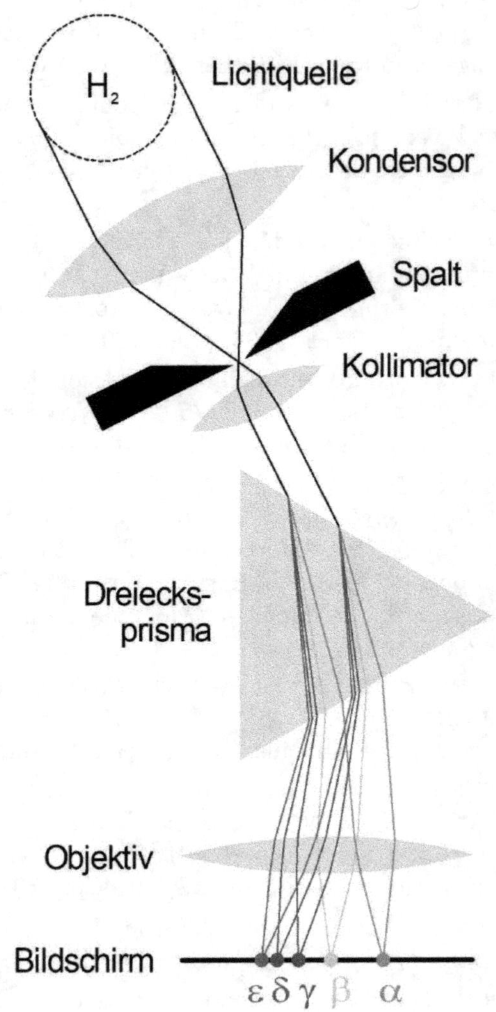

Abbildung 2.1:
Die Gewinnung der fünf sichtbaren Spektrallinien H_α, H_β, H_γ, H_δ und H_ε der Balmer-Serie mit einem refraktiven Spektrometer:

Ein von einer Lichtquelle mit leuchtendem Wasserstoff mit Hilfe eines Kondensors beleuchteter Spalt wird durch einen Kollimator nach Unendlich abgebildet. Durch die Dispersion des Dreiecksprismas werden die kurzwelligen (violetten) Anteile des Lichtes stärker gebrochen als die langwelligen (roten) Anteile. Ein dahinter befindliches Objektiv bildet die kollimierten Lichtstrahlen auf einen Bildschirm ab, so dass dort an verschiedenen Stellen verschiedenfarbige Bilder des Spaltes zu sehen sind, die den Emissionslinien der Lichtquelle entsprechen.

(Grafik: Markus Bautsch)

Die von Johann Jakob Balmer gefundenen Zahlenverhältnisse bei den Spektrallinien des Wasserstoffs

Markus Bautsch (Berlin)

Abstract: The arithmetical ratios for the spectral lines of hydrogen found by Johann Jakob Balmer

The absorption lines in the solar spectrum were already observed and described at the beginning of the 19th century. The discovery of the numerical ratios of the wavelengths of the spectral lines of hydrogen by Johann Jakob Balmer (1825–1898) in 1884 astonished and perplexed the world in equal measure.

Numerous scientists have made important contributions to research into these spectral lines. It was not until 1926 that the mystery of the Balmer series was solved with the help of quantum mechanics.

The mathematician and composer Hans Sommer (1837–1922) befriended the young composer Richard Strauss (1864–1949) in the early 1890s and apparently told him about it. Because in 1893, Strauss began a composition, in which the pitches of a theme correspond to the frequencies of the visible spectral lines of the Balmer series.

Zusammenfassung

Die Absorptionslinien im Sonnenspektrum wurden bereits zu Beginn des 19. Jahrhunderts beobachtet und ausführlich beschrieben. Die Entdeckung der Zahlenverhältnisse bei den Wellenlängen der Spektrallinien des Wasserstoffs durch Johann Jakob Balmer (1825–1898) versetzte die Welt 1884 jedoch gleichermaßen in Erstaunen und Ratlosigkeit.

Zahlreiche Wissenschaftler haben bis ins 20. Jahrhundert hinein wichtige Beiträge zur Erforschung und Erklärung dieser Spektrallinien geliefert, und erst 1926 konnte das Rätsel der Balmer-Serie mit Hilfe der Quantenmechanik gelöst werden.

Der Mathematiker und Komponist Hans Sommer (1837–1922) freundete sich Anfang der 1890er Jahre mit dem jungen Komponisten Richard Strauss (1864–1949) an und erzählte ihm offensichtlich davon, denn 1893 begann Strauss mit einer Komposition, bei der die Tonhöhen eines Motivs den Frequenzen der sichtbaren Spektrallinien der Balmer-Serie entsprechen.

2.1 Spektrallinien

2.1.1 Die Entdeckung der Absorptionslinien im Sonnenspektrum

Der englische Arzt, Physiker und Chemiker William Hyde Wollaston (1766–1828) war nicht nur Entdecker der beiden Edelmetalle Palladium und Rhodium. Am Anfang des 19. Jahrhunderts betrachtete er aus einer Entfernung von gut drei Metern durch einen 1,3 mm breiten schmalen Spalt in ein Prisma aus Flintglas einfallendes Sonnenlicht. Hierbei bemerkte er im visuellen Spektrum drei deutliche dunkle Streifen. In seinem Kapitel *Über eine Methode zur Untersuchung refraktiver und dispersiver Kräfte bei prismatischer Reflexion* schrieb er im Juni 1802 folgendes:

> „*By employing a very narrow pencil of light, 4 primary divisions of the prismatic spectrum may be seen, with a degree of distinctness that, I believe, has not been described nor observed before.*“[1]

Diese dunklen Streifen hielt er für Trennlinien zwischen den natürlichen Farben Rot, Gelbgrün, Blau und Violett. Zusammen mit den durch Temperaturmessung nachweisbaren infraroten und durch Fluoreszenz nachweisbaren ultravioletten Strahlen zählte Wollaston in seinem Beitrag insgesamt sechs Arten von Licht, deren Grenzen er mit den Buchstaben A bis E kennzeichnete. 1820 wurde Wollaston auswärtiges Mitglied der Bayerischen Akademie der Wissenschaften.

2.1.2 Spektroskopische Messungen

1814 wurden zahlreiche solcher dunklen Linien unabhängig von Wollastons Beobachtungen vom deutschen Optiker Joseph von Fraunhofer (1787–1826) ent-

1 Wollaston 1802, S. 378–380.

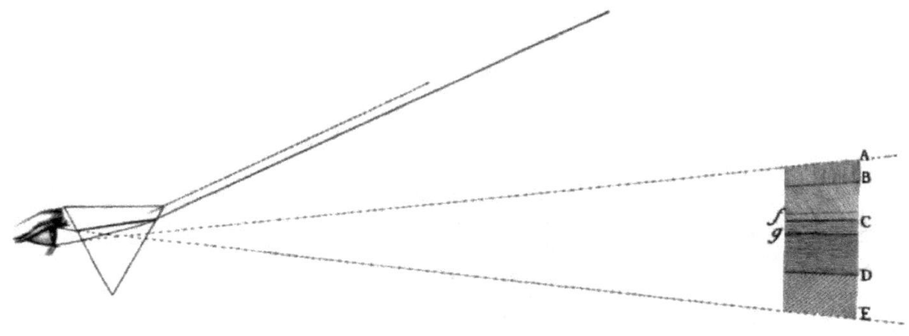

Abbildung 2.2:
Originale Skizze von Wollaston
mit dem durch ein Dreiecksprisma aufgefächertes Sonnenspektrum
bei der visuellen Betrachtung durch einen Spalt

(Wollaston 1802, Plate XIV. Fig. 3)

deckt, systematisch untersucht, ebenfalls mit Kennbuchstaben versehen und dokumentiert. Ihm zu Ehren werden sie bis heute Fraunhofer-Linien genannt.

Eine quantitative Bestimmung der Lichtwellenlängen war mit Hilfe des von ihm verwendeten Prismas allerdings nicht möglich. Die präzise Messung der Brechungswinkel führte er mit dem von ihm erfundenen Spektroskop durch, bei dem das zu untersuchende Licht durch einen schmalen senkrechten Spalt und ein Dreiecksprisma geschickt und dahinter mit einem auf einem Azimutalkreis befestigten Fernrohr mit beobachtet wurde.

1817 legte Fraunhofer seine Abhandlung *Bestimmung des Brechungs- und Farbenzerstreuungsvermögens verschiedener Glasarten in Bezug auf die Vervollkommnung achromatischer Fernröhre*[2] der Bayerischen Akademie der Wissenschaften vor, von der er im selben Jahr auf Antrag des Königlichen Hofastronomen Johann Georg von Soldner (1776–1833) zu einem korrespondierenden Mitglied gewählt wurde. 1821 wurde Fraunhofer schließlich als außerordentliches Mitglied und 1823 als ordentliches Mitglied in die Akademie aufgenommen.

2.1.3 Künstliche Lichtquellen mit Spektrallinien

1857 gelang es dem Glasbläser Heinrich Geißler (1814–1879) Metalldrähte luftdicht in Glasröhren einzuschmelzen und diese zu evakuieren. Mit solchen Niederdruck-

2 Fraunhofer 1814/1815.

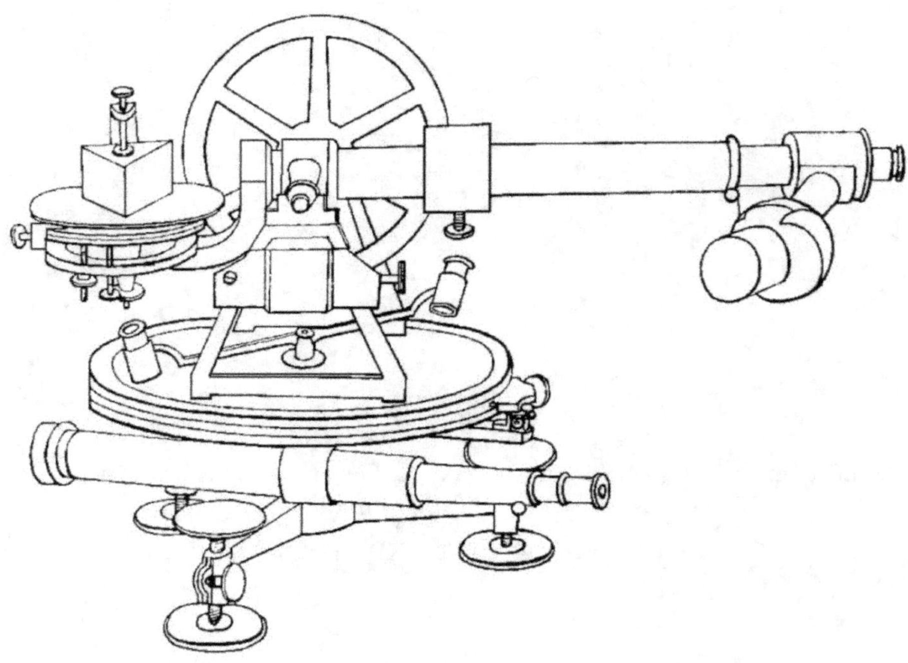

Abbildung 2.3:
Fraunhofers originale Skizze mit seinem Spektrometer und einem 60°-Dreiecksprisma.
Mit dem Beobachtungsfernrohr wurde ein 0,5 mm schmaler, vertikal ausgerichteter und von hinten beleuchteter Spalt aus einer Entfernung von gut sieben Metern durch das Prisma hindurch beobachtet, so dass die Spektrallinien der Lichtquelle erkennbar wurden.

(Fraunhofer 1814/1815)

Kaltkathoden-Gasentladungsröhren, die oft Geißler-Röhren genannt werden, konnte fürderhin die Lichtemission von verschiedenen Gasen untersucht und analysiert werden.[3]

Gustav Robert Kirchhoff (1824–1887) und Robert Wilhelm Bunsen (1811–1899) entdeckten 1859 in spektroskopisch beobachteten Gasflammen, dass die Spektrallinien durch chemische Elemente verursacht werden und dass sie kennzeichnend für diese sind.[4] Durch diese Erkenntnis konnte nicht nur die Ursache

3 Schellen 1859.
4 Kirchhoff & Bunsen 1860.

der vielen Fraunhofer-Linien erklärt werden, sondern 1861 konnten die beiden Forscher mit ihrer Methode sogar die von ihnen neu entdeckten Alkalimetalle Cäsium und Rubidium identifizieren. Mit dieser Art der Spektralanalyse war nun auch eine wesentliche Grundlage für die moderne Astronomie gegeben.

2.1.4 Wellenlängenmessung mit Transmissionsgittern

Der schwedische Astronom und Physiker sowie Mitbegründer der Astrospektroskopie Anders Jonas Ångström (1814–1874) konnte mit Hilfe eines Spektroskops, das mit einem Transmissionsgitter ausgestattet war, die Wellenlängen der Spektrallinien präzise bestimmen. 1868 konnte er die vier Wellenlängen der bekannten und sichtbaren Wasserstofflinien angeben:[5]

$$H_\alpha \ 656{,}21\,\mathrm{nm}\ (\mathrm{rot}),\ H_\beta \ 486{,}074\,\mathrm{nm}\ (\mathrm{t\ddot{u}rkis}),$$
$$H_\gamma \ 434{,}01\,\mathrm{nm}\ (\mathrm{blau}),\ H_\delta \ 410{,}12\,\mathrm{nm}\ (\mathrm{violett})$$

2.1.5 Nachweis von Spektrallinien im Sternenlicht

In der Folge wurde bekannt, dass auch bei Fixsternen diese vier Wasserstofflinien auftreten und diese somit ein ähnliches Emissionsverhalten zeigen wie unsere Sonne. Bis zum Jahr 1867 konnte der italienische Jesuit und Astronom Angelo Secchi (1818–1878) bereits drei Spektralklassen festlegen, in die er die ungefähr 500 von ihm untersuchten Fixsterne einteilen konnte.[6]

Die Fraunhofer-H-Linie bei 397 nm (violett) konnte 1879 vom britischen Astronomen und Physiker William Huggins (1824–1910) dank der von ihm auch für ultraviolettes Licht einsetzbaren optischen Instrumente mit Silberspiegeln und Quarzprismen im Sternenspektrum der hellen und heißen Sterne Sirius und Wega ebenfalls nachgewiesen werden.[7] Diese H-Linie war bis dahin immer dem chemischen Element Calcium zugeordnet worden.

2.1.6 Weitere Spektrallinien des Wasserstoffs

Der deutsche Photochemiker Hermann Wilhelm Vogel (1834–1898) fand im Jahr 1879 in den Aufnahmen einer wasserstoffhaltigen Geißler-Röhre neben

5 Ångström 1868.
6 Secchi 1867.
7 Huggins 1880.

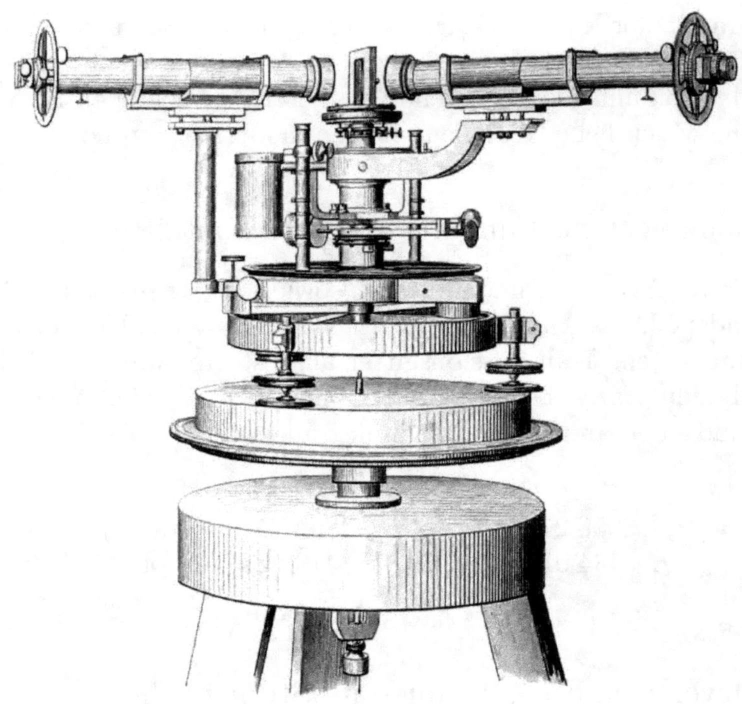

Abbildung 2.4:
Ångström Spektrometer, Pistor & Martins, Berlin

Das von Anders Jonas Ångström verwendete Spektrometer, das anstelle eines Dreiecksprismas mit einem Transmissionsgitter und zudem mit einem Kollimatorrohr ausgestattet ist. Das Licht wird nicht an den Grenzflächen eines optischen Mediums in Abhängigkeit von seiner Wellenlänge gebrochen, sondern an den Gitterlinien gebeugt, wodurch sich ein direkter geometrischer Zusammenhang zwischen der Lichtwellenlänge und jeweiligen Richtung der Beugungsordnungen ergibt.

(Ångström 1868, Frontispiz)

den vier bekannten Spektrallinien des Wasserstoffs mehrere weitere Spektrallinien im Violetten und im Ultravioletten:[8]

396,9 nm (violett) / 388,7 nm / 383,4 nm / 379,5 nm / 376,75 nm (ultraviolett)

8 Vogel, H. W. 1880.

H. W. Vogel mutmaßte, dass es sich bei der violetten Spektrallinie bei der Wellenlänge von knapp 397 Nanometern um eine fünfte Wasserstofflinie (H_ε) handeln dürfte, wovon er Huggins schnell überzeugen konnte. Diese Spektrallinie unterscheidet sich in der Wellenlänge nur um 0,16 nm von der Spektrallinie Ca II H des ionisierten Calciums, von der sie zuvor nicht unterschieden worden war.

Diese wichtigen Ergebnisse von Vogel wurden 1880 von Hermann von Helmholtz (1821–1894) bei einer Sitzung der Königlichen Akademie der Wissenschaften in Berlin vorgestellt und somit einem großen Kreis von Fachleuten mitgeteilt. In der folgenden Tabelle sind die fünf sichtbaren Spektrallinien des Wasserstoffs aufgeführt.

Tabelle 2.1:
Die fünf sichtbaren Spektrallinien des Wasserstoffs

Wellenlänge in nm	Spektrallinie	Fraunhofers Bezeichnung	Farbe
656,2	H_α	C-Linie	Rot
486,1	H_β	F-Linie	Blaugrün
434,0	H_γ	f-Linie vor G	Blau
410,1	H_δ	h-Linie vor H	Violett
396,8	H_ε	∼H-Linie	Violett

Die Einteilung der Fixsterne in Spektralklassen von Secchi wurde von Hermann Carl Vogel (1841–1907) 1874 modifiziert. Die Harvard-Klassifikation mit den 7 Spektraltypen (O, B, A, F, G, K, M) wurde im wesentlichen von Annie Jump Cannon (1863–1941) 1901 begonnen und 1918 bis 1923 geschaffen. 1922 wurde sie von der *International Astronomical Union* (IAU) anerkannt und ist bis heute in Verwendung (später erweitert um einige spezielle Spektralklassen).[9]

H. W. Vogel führte einen der entscheidenden Begründer der Spektroskopie, Heinrich Gustav Johannes Kayser (1853–1940), in die damals noch junge Technik der Photographie ein. Kayser baute ab 1894 in Bonn ein modernes sowie das erste vollständig der Spektroskopie gewidmete Institut auf. Ab 1900 gab er das umfassende und damals sehr bedeutende *Handbuch der Spektroskopie* her-

9 Wolfschmidt 2022, hier S. 93–97, Kapitel 2.4.3: Auf dem Weg zur Klassifikation von Sternspektren.

aus (siehe Kapitel 3, S. 103), das für die weitere Entwicklung der Atomphysik eine grundlegende Rolle spielte.

2.2 Die Balmer-Serie

Der Schweizer Mathematiker und Physiker Johann Jakob Balmer (1825–1898) war auch an der Philologie und an Architektur interessiert. Seine Dissertation hatte er über die Zykloide geschrieben, und er galt als Fachmann für Darstellende Geometrie. Er war verheiratet, hatte sechs Kinder und betätigte sich in Basel als Lehrer, als Privatdozent und als Schulinspektor. Er war als Mitglied im Basler Grossen Rat und betätigte sich außerdem als Armenpfleger sowie als Kirchenvorstand.

1884 konnte Balmer empirisch zeigen, dass für die von Anders Jonas Ångström präzise experimentell bestimmten vier Wellenlängen der Spektrallinien des Wasserstoffs ein einfacher arithmetischer Zusammengang mit einem ganzzahligen Koeffizienten n angegeben werden kann.

Mit Hilfe der von ihm *Grundzahl des Wasserstoffs* genannten Größe a = 364,51 nm ermittelte er den folgenden arithmetischen Zusammengang zwischen den ganzen Zahlen n = 3, 4, 5 und 6 und den Wellenlängen λ_n der ersten vier Fraunhofer-Linien des Wasserstoffs:

$$\lambda_n = a \cdot \frac{n^2}{n^2 - 4} \quad \text{für } n > 2$$

In seiner *Notiz über die Spektrallinien des Wasserstoffes* schreibt Balmer 1885:

> *„Es sind besonders die numerischen Verhältnisse der Wellenlängen der ersten vier Wasserstofflinien, welche die Aufmerksamkeit reizen und fesseln. Die Verhältnisse dieser Wellenlängen lassen sich nämlich überraschend genau durch kleine Zahlen ausdrücken.*
> *Der Unterschied zwischen den berechneten und beobachteten Wellenlängen ist so klein, dass die Übereinstimmung im höchsten Grade überraschen muss.“*[10]

Balmer bemerkte ferner, dass auch die damals erst seit wenigen Jahren bekannte fünfte Spektrallinie des Wasserstoffs im Violetten außerordentlich gut mit der von ihm gefundenen Zahlenfolge im Einklang steht:

10 Balmer 1885, S. 551.

Abbildung 2.5:
Johann Jakob Balmer (1825–1898) in den 1880er Jahren

(Creative Commons)

„Aus diesen Vergleichungen ergibt sich [...], dass die Formel auch für die fünfte [...] Wasserstofflinie zutrifft.“[11]

Im Weiteren ergab sich, dass die vier weiteren, im Ultravioletten liegenden Spektrallinien ebenfalls perfekt zu der von ihm gefundenen Zahlenfolge passen:

Aus diesen Übereinstimmungen schlussfolgerte Balmer zutreffend, dass diese Spektrallinien bei den Wellenlängen für n gleich 7 bis 11 und gegebenenfalls

11 Balmer 1885, S. 556.

Tabelle 2.2:
Die Wellenlängen der neun ersten Spektrallinien der Balmer-Serie

n	$\frac{n^2}{n^2-4}$	λ_n in nm
3	9/5	656,2
4	4/3	486,1
5	21/21	434,0
6	9/8	410,1
7	49/45	397,0
8	16/15	388,9
9	81/77	383,5
10	25/24	379,8
11	121/117	377,0

sogar noch weitere, noch kurzwelligere Spektrallinien dem chemischen Element Wasserstoff zugeordnet werden können.

2.2.1 Verallgemeinerung durch die Rydberg-Formel

Ende 1888 stellte der schwedische Physiker Johannes Rydberg (1854–1919) die verallgemeinerte *Rydberg-Formel* mit zwei ganzzahligen Koeffizienten m und n vor.[12] In Anlehnung an die von Balmer gewählten Schreibweise beschreibt sie mit der Rydberg-Konstante genannten Zahl $R = \frac{4}{a}$ die Wellenlängen in der folgenden Form:

$$\lambda_{m,n} = \frac{a}{4 \cdot \left(\frac{1}{m^2} - \frac{1}{n^2}\right)} = \frac{1}{R \cdot \left(\frac{1}{m^2} - \frac{1}{n^2}\right)}$$

Die Rydberg-Konstante ist heute die am genauesten gemessene Naturkonstante überhaupt. Die Balmer-Serie ergibt sich hieraus für m=2, wovon man sich durch Einsetzen leicht überzeugen kann.

Im selben Jahr bewies der deutsche Physiker Heinrich Hertz (1857–1894) durch die Erzeugung elektromagnetischer Wellen die Vermutung, dass auch Licht eine elektromagnetische Welle mit der Wellenlänge λ und der Frequenz f ist, und dass sich alle elektromagnetischen Wellen mit der Lichtgeschwindigkeit

12 Rydberg 1889.

c ausbreiten:[13]

$$\lambda = \frac{c}{f}$$

2.3 Musikalischer Exkurs

Der deutsche Mathematiker und Komponist Hans Zincken, genannt Sommer (1837–1922) veröffentlichte als 21-Jähriger seine Dissertation *„Zur Bestimmung der Brechungsverhältnisse“*, in der er die spektrale Zerlegung von Licht untersuchte. Als Balmer seine Serie 1884 entdeckte, beschäftigte er sich in Braunschweig ausführlich mit angewandter Optik und war sicherlich über die Entdeckung Balmers informiert.

Anfang der 1890er Jahre freundete Sommer sich in Weimar mit dem jungen Komponisten Richard Strauss (1864–1949) an. Mit großer Wahrscheinlichkeit haben sich die beiden auch über die rätselhafte Balmer-Serie und die Übertragung derer Schwingungszahlen auf die des Schalls ausgetauscht, denn 1893 begann Strauss mit der Komposition seiner Sinfonischen Dichtung *Till Eulenspiegels lustige Streiche.* Das schalkhafte Eingangsmotiv für Till Eulenspiegel besteht aus den fünf Tönen c–f–g–gis–a, deren Tonhöhen hoch signifikant den Frequenzen der fünf ersten und sichtbaren Spektrallinien der Balmer-Serie entsprechen.

Abbildung 2.6:
Die fünf Töne c–f–g–gis–a des Till-Eulenspiegel-Motivs von Richard Strauss

(Creative Commons)

Die Information über die gleichen Zahlenverhältnisse bei der Balmer-Serie und bei dieser Tonfolge verdankt der Autor Prof. Gerd Koppelmann (1929–1992), der sie in den 1980er Jahren mit Bezug auf die Erzähltradition seiner Physik-Professoren während einer physikalischen Lehrveranstaltung an der Technischen Universität Berlin erwähnt hatte.

13 Hertz 1887.

Strauss legte für sein musikalisches Motiv die physikalische Stimmung auf dem Grundton C mit 256 Hertz ($256 = 2^8$) zugrunde, in welcher der Ton C in allen Oktaven als ganzzahlige Potenz von Zwei dargestellt werden kann. Der damals in der entsprechenden Pariser Stimmung verwendete Stimmton A hatte demzufolge eine Frequenz von nur 431 Hertz.

Auch bei der kurz darauf komponierten Sinfonischen Dichtung *„Also sprach Zarathustra"* bediente sich Strauss einer physikalischen Zahlenfolge. Für diese Komposition wählte er für das Eingangsmotiv die ersten fünf Töne der Obertonreihe von Blas- und Saiteninstrumenten und wählte als Bezugstonhöhe erneut die physikalische Stimmung auf dem Grundton C, so dass sich die folgende Tonfolge ergab: C–c–g–c′–e′.

2.4 Quantenphysik

2.4.1 Die Naturkonstante h

Wilhelm Wien (1864–1928) hatte 1896 das *Wiensche Strahlungsgesetz* für die spektrale Lichtemission von Schwarzen Körpern gefunden.[14] Seine Ergebnisse waren in Übereinstimmung mit den Beobachtungen von Friedrich Paschen (1865–1947), dessen Tante Adele Henriette Dorothea Caroline Paschen (1848–1908) mit dem Astronomen Wilhelm Foerster (1832–1921) verheiratet war. Beide Physiker untersuchten bei der thermischen Strahlung empirisch die Abhängigkeit der Strahlungsdichte von der Wellenlänge.

Der theoretische Physiker Max Planck (1858–1947) veröffentlichte 1900 das mit klassischer Physik nicht erklärbare und von ihm ebenfalls nur empirisch gefundene *Plancksche Strahlungsgesetz*, das die Hohlraumstrahlung noch genauer beschreibt. Zur Vereinfachung der Formulierung führte er mit dem *Planckschen Wirkungsquantum h* eine „Hilfsgröße" ein, die sich auch mit den Konstanten des Wienschen Strahlungsgesetzes ausdrücken lässt.[15] 1918 erhielt Planck für diese bahnbrechende Arbeit den Physik-Nobelpreis.

Heinrich Hertz (1857–1894) war bereits 1886 aufgefallen, dass die ultraviolette Strahlung eines Funkens die Länge einer zweiten Funkenstrecke vergrößert.[16] Wilhelm Hallwachs (1859–1922) entdeckte 1888 den *Hallwachs-Effekt*, bei dem sich zeigt, dass sich eine Metallplatte bei Bestrahlung mit ultraviolettem Licht auflädt.[17] Im Jahr 1899 führte Philipp Lenard (1862–1947) Untersuchungen im

14 Wien 1896.
15 Planck 1900.
16 Hertz 1886.
17 Hallwachs 1888.

Hochvakuum durch und stellte fest, dass zwar die Anzahl der durch ultraviolettes Licht an einer Kathode ausgelösten Elektronen mit der Lichtintensität ansteigt, dass aber die kinetische Energie dieser Elektronen davon völlig unabhängig ist.[18]

Auf Basis dieser experimentellen Befunde entdeckte Albert Einstein (1879–1955) 1905 die Teilchennatur des Lichtes,[19] wofür er 1921 mit dem Physik-Nobelpreis belohnt wurde. Photonen sind demnach Teilchen mit einer Energie E, die proportional zur Schwingungsfrequenz f der elektromagnetischen Welle und somit umgekehrt proportional zu deren Wellenlänge λ ist. Die Proportionalitätskonstante ist das Plancksche Wirkungsquantum h:

$$E = h \cdot f = h\frac{c}{\lambda}$$

Robert Andrews Millikan (1886–1953) konnte zehn Jahre später mit der von ihm entwickelten Gegenfeldmethode experimentell nachweisen, dass die Proportionalitätskonstante h aus der Einsteinschen Formel mit dem Planckschen Wirkungsquantum h identisch ist.[20]

2.4.2 Weitere Wasserstoffserien

Nach Einsteins Entdeckung wurden in Übereinstimmung mit der verallgemeinerten Rydberg-Formel neben der Balmer-Serie mit m=2 weitere Serien von Spektrallinien gefunden.

Der US-amerikanische Physiker Theodore Lyman (1874–1954) entdeckte 1906 die für die astronomische Beobachtung wichtige Lyman-Serie im Ultravioletten für m=1.[21]

Danach wurden weitere Wasserstoffserien im Infraroten gefunden:

- 1908 entdeckte der deutsche Physiker Friedrich Paschen (1865–1947) die Paschen-Serie für m=3.[22]
- Die Brackett-Serie für m=4 wurde 1922 von dem US-amerikanischen Astronomen Frederick Sumner Brackett (1896–1988) entdeckt.[23]
- Die Pfund-Serie für m=5 wurde im Jahr 1924 von dem US-amerikanischen Physiker August Herman Pfund (1879–1949) entdeckt.[24]

18 Lenard 1900.
19 Einstein 1905.
20 Millikan 1916.
21 Lyman 1906.
22 Paschen 1908.
23 Brackett 1922.
24 Pfund 1924.

- Die sehr langwellige Humphreys-Serie für m=6 wurde schließlich im Jahr 1952 von dem US-amerikanischen Physiker Curtis Judson Humphreys (1898–1986) gefunden.[25]

2.4.3 Das Bohrsche Atommodell

Der dänische Physiker Niels Bohr (1885–1962) entwickelte 1913 das Bohrsche Atommodell mit Quantensprüngen zwischen den Energieniveaus der Elektronen in Atomen.[26] Hierfür formulierte er die *Bohrschen Postulate.* Danach befinden sich die Elektronen in den stationären Zuständen des Atoms auf ausgewählten Bahnen, ohne dabei Energie zu verlieren und elektromagnetische Strahlung zu erzeugen. Diese stationären Bahnen sind diskreten Energiezuständen zugeordnet, und der dazugehörige Bahndrehimpuls ist stets ein ganzzahliges Vielfaches von $\hbar = \frac{h}{2\pi}$. Die Elektronen können zwischen diesen diskreten Energiezuständen Quantensprünge vollziehen, und dabei werden Photonen absorbiert oder emittiert, deren Energie exakt der Energiedifferenz dieser Energiezustände entspricht.

Für diesen elementaren Schritt zum tieferen Verständnis der Vorgänge in der Atomhülle bekam Bohr 1922 den Nobelpreis für Physik zugesprochen.

2.4.4 Die Schrödinger-Gleichung

Erst 1926 konnte vom österreichischen Physiker Erwin Schrödinger (1887–1961) mit Hilfe der Quantenmechanik das Rätsel der Balmer-Serie gelöst werden. Er wendete den zeitabhängigen *Hamilton-Operator* $H(t)$ auf die Wellenfunktion $\Psi(t)$ an, um mit der nach ihm benannten Schrödinger-Gleichung die Ausbreitung von Materiewellen berechnen zu können.[27] 1933 erhielt Schrödinger für seine Leistung den Physik-Nobelpreis.

Die Lösung der Schrödinger-Gleichung liefert in Übereinstimmung mit der Rydberg-Formel und somit auch mit der Balmer-Formel alle diskreten Energieniveaus des Elektrons in der Atomhülle eines Wasserstoffatoms. Unter Berücksichtigung der kinetischen Energie des Elektrons und der potenziellen Energie im elektrischen Feld zwischen dem positiv geladenen Atomkern und dem negativ geladenen Elektron verhalten sich die diskreten Energieniveaus E_n der Elektronen im Wasserstoffatom danach wie die Kehrwerte der Quadrate ganzer

25 Humphreys 1953.
26 Bohr 1913.
27 Schrödinger 1926.

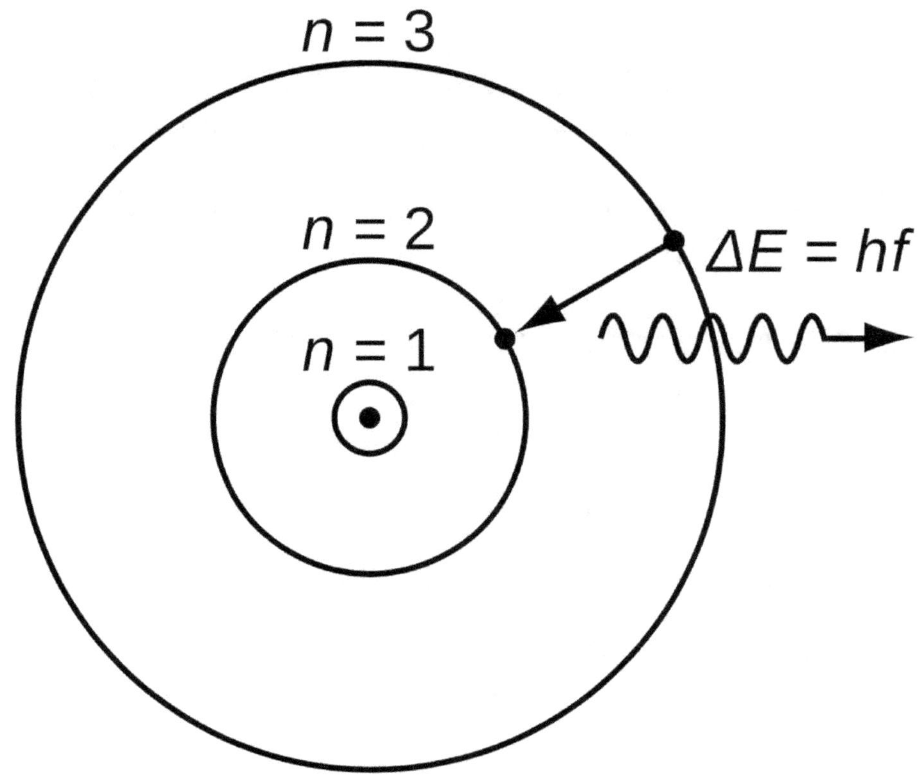

Abbildung 2.7:
Bohrsches Atommodell für Wasserstoff

Das Bohrsche Atommodell für Wasserstoff mit einem einfach positiv geladenen Proton im Kern. Die Energiedifferenz ΔE beim Übergang eines einfach negativ geladenen Elektrons zwischen den Energieniveaus E_3 und E_2 ist proportional zur Frequenz f des emittierten Photons. Die Proportionalitätskonstante ist das Plancksche Wirkungsquantum h.

(Grafik: Markus Bautsch)

Zahlen n (mit $n = 1$):

$$E_n = \frac{E_1}{n^2} = -hc \cdot \frac{4}{a \cdot n^2} = -hc \cdot \frac{R}{n^2} = -\frac{hc}{e} \cdot \frac{R}{n^2} \text{ [in eV]}$$

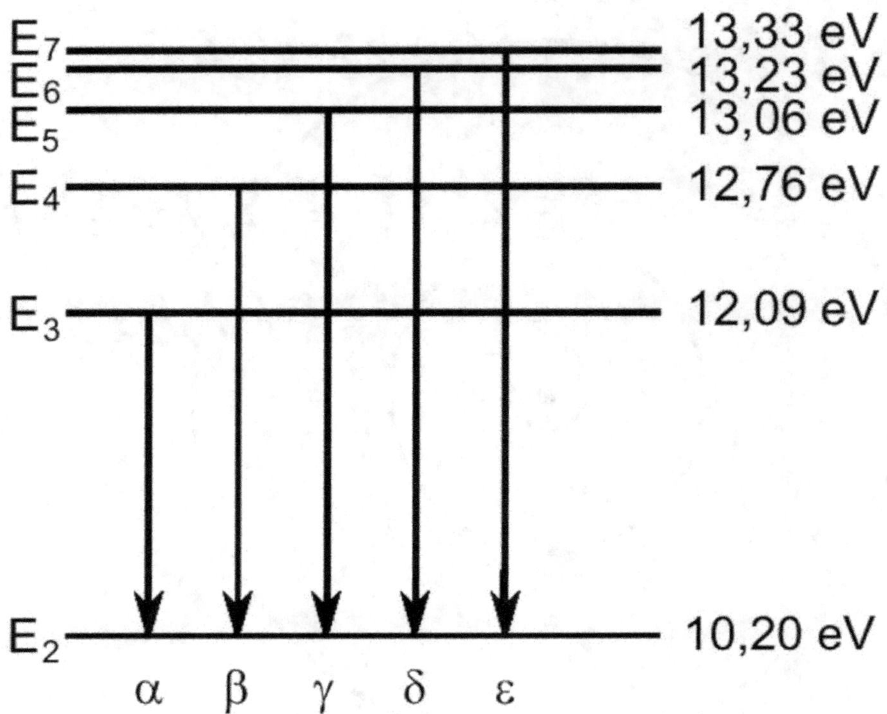

Abbildung 2.8:
Die Energieniveaus E_2 bis E_7 für die fünf ersten Linien der Balmer-Serie in Elektronenvolt (eV). Die Energie $E_1 = -2,1798710^{-18}$ J = -13,6057 eV entspricht dem Grundzustand des Elektrons, auf den alle in dieser Darstellung angegebenen Energien bezogen sind.

(Grafik: Markus Bautsch)

2.5 Wasserstofflinien in der Astronomie

Absorptionslinien sind praktisch bei allen Sternen zu beobachten. Bei Sternen mit einer Oberflächentemperatur von knapp 10.000 Kelvin (Spektralklasse A) sind sie am stärksten ausgeprägt. Emissionslinien können in den Spektren vieler Galaxien, aktiver galaktischer Kerne, interstellarer Wolken und planetarischer Nebel nachgewiesen werden.

Die Wasserstofflinie H_α hat eine wichtige Bedeutung bei der Sonnenbeobachtung im roten Licht. H-alpha-Teleskope sind mit Fabry-Pérot-Interferometern oder dichrotischen Spiegeln ausgestattet, die das Licht in einem nur sehr schmalen Wellenlängenbereich bei der H_α-Linie hindurchlassen (656,28 Nanometer), so dass Sonnenflecke und die Strukturen der Chromosphäre erheblich besser beobachtet werden können.

Mit Hilfe der ersten Lyman-Linie (Lyman α) mit der Wellenlänge

$$\lambda_{1,2}{=}121{,}50\,\mathrm{nm}$$

wird in der modernen Astrophysik sowohl die Rotverschiebung weit entfernter Galaxien und Quasare als auch die weiträumige Verteilung von Wasserstoff im Universum untersucht. Die Rotverschiebung z ist über die Differenz $\Delta\lambda$ der bei der Beobachtung der Lichtwellenlänge λ eines astronomischen Objekts mit der Wellenlänge λ_0 definiert:

$$z = \frac{\Delta\lambda}{\lambda_0} = \frac{\lambda - \lambda_0}{\lambda_0}$$

Tabelle 2.3:
Rotverschiebung z und Wellenlänge

Objekt	Rotverschiebung z	Wellenlänge λ (Lyman α)
Sonne	$2{,}1 \cdot 10^{-6}$	121,5 nm
Quasar 3C 9 (1965)	2,018	367 nm
Supernova SN 1000+0216	3,9	595 nm
Quasar QSO J0313-1806	7,64	1050 nm
Galaxie JADES-GS-z13-0	13,2	1725 nm
Hintergrundstrahlung (2,7 K)	1089	1,3 mm
Schwarzes Loch	∞	∞

Die Ursache für die Rotverschiebung kann sowohl an der Wirkung der Gravitation von sehr großen Massen (allgemeine Relativitätstheorie) als auch an der Relativgeschwindigkeit zwischen Beobachter und Objekt (spezielle Relativitätstheorie) liegen. Sofern sich astronomische Objekte, wie zum Beispiel Doppelsterne, Exoplaneten oder der Wasserstoff in den Akkretionsscheiben von Neutronensternen und Schwarzen Löchern in Beobachtungsrichtung bewegen, kann durch die entsprechende Rot- oder Blauverschiebung deren Bewegung nachgewiesen und gegebenenfalls sogar die Relativgeschwindigkeit bestimmt werden. Durch den Vergleich der Rotverschiebung von astronomischen Objekten mit ähnlichen Bewegungen, wie beispielsweise in Sternhaufen, in Galaxienhaufen oder bei Trümmern von Kollisionen kann die Zusammengehörigkeit dieser Objekte ermittelt werden.

2.6 Literatur

ÅNGSTRÖM, ANDERS JONAS: *Recherches sur le Spectre Solaire. Spectre normal du Soleil – Atlas de six planches de Robert Thalén (1827–1905).* Uppsala: Verlag W. Schultz 1868.

BALMER, JOHANN JAKOB: Notiz über die Spektrallinien des Wasserstoffes. In: *Verhandlungen der Naturforschenden Gesellschaft in Basel*, Band 7. Basel: H. Georg's Verlag 1885, S. 548–560. Siehe auch: BALMER, JOHANN JAKOB: Notiz über die Spektrallinien des Wasserstoffes. In: *Annalen der Physik und Chemie*, Band **261** = Neue Folge Band 25 (1885), Nr. 5, S. 80–87.

BOHR, NIELS: On the Constitution of Atoms and Molecules. In: *Philosophical Magazine* **26** (1913), Teil I, S. 1–25, Teil II, S. 476–502, Teil III, S. 857–875.

BRACKETT, FREDERICK SUMNER: Visible and Infra-Red Radiation of Hydrogen. In: *Astrophysical Journal* **56** (1922), S. 154–161.

EINSTEIN, ALBERT: Ueber einen die Erzeugung und Verwandlung des Lichtes betreffenden heuristischen Gesichtspunkt. In: *Annalen der Physik* **322** (1905), Nr. 6, S. 132–148.

FRAUNHOFER, JOSEPH: Bestimmung des Brechungs- und Farbzerstreuungs-Vermögens verschiedener Glasarten, in bezug auf die Vervollkommnung achromatischer Fernröhre. In: *Denkschriften der Königlichen Akademie der Wissenschaften zu München für die Jahre 1814 und 1815*, Band V. München 1815, S. 193–226.

HALLWACHS, WILHELM: Ueber den Einfluss des Lichtes auf electrostatisch geladene Koerper. In: *Annalen der Physik und Chemie* **269** (1888), Nr. 2, S. 301–312.

HEINRICH SCHELLEN: Die Inductions-Electrizität, zweiter Artikel. In: *Natur und Offenbarung – Organ zur Vermittlung zwischen Naturforschung und Glauben für Gebildete aller Stände*, fünfter Band. Münster: Aschendorff'sche Buchhandlung 1859, S. 337–355.

HERTZ, HEINRICH: Ueber sehr schnelle electrische Schwingungen. In: *Annalen der Physik und Chemie* **267** (1887), Nr. 7, S. 421–448.

HERTZ, HEINRICH: Ueber einen Einfluss des ultravioletten Lichtes auf die electrische Entladung. In: *Annalen der Physik und Chemie* **267** (1887), Nr. 8, S. 983–1000.

HUGGINS, WILLIAM: On the Photographic Spectra of Stars. In: *Philosophical Transactions of the Royal Society of London* **171** (1880), S. 669–690.

Humphreys, Curtis Judson: The Sixth Series in the Spectrum of Atomic Hydrogen. In: *Journal of Research of the National Bureau of Standards* **50** (1953), Nr. 1, S. 1–6.

Kirchhoff, Gustav & Robert Bunsen: Chemische Analyse durch Spectralbeobachtungen. In: *Annalen der Physik und Chemie* **110** (1860), Nr. 6, S. 161–189.

Lenard, Philipp: Erzeugung von Kathodenstrahlen durch ultraviolettes Licht. In: *Annalen der Physik* **307** (1900), Nr. 6, S. 359–375.

Lyman, Theodore: The Spectrum of Hydrogen in the Region of Extremely Short Wave-Lengths. In: *Astrophysical Journal* **23** (1906), S. 181–210.

Millikan, Robert Andrews: A Direct Photoelectric Determination of Planck's h. In: *Physical Review* **7** (1916), Nr. 3, S. 355–388.

Paschen, Friedrich: Zur Kenntnis ultraroter Linienspektra. In: *Annalen der Physik* (4) **27**, Band 332 der gesamten Reihe, (1908), Nr. 13, S. 537–570.

Pfund, August Herman: The Emission of Nitrogen and Hydrogen in the Infrared. In: *Journal of the Optical Society of America* **9** (1924), Nr. 3, S. 193–196.

Planck, Max: Zur Theorie des Gesetzes der Energieverteilung im Normalspectrum. In: *Verhandlungen der Deutschen Physikalischen Gesellschaft*, Nr. 17. Berlin 1900, S. 237–254.

Rydberg, Johannes Robert: Recherches sur la constitution des spectres d'émission des éléments chimiques. In: *Kongliga Svenska Vetenskaps-Akademiens Handlingar* **23** (1889), Nr. 11, S. 1–177.

Schrödinger, Erwin: Quantisierung als Eigenwertproblem, Teil I–IV. In: *Annalen der Physik* **79–81** (1926), Band 79, Teil I, S. 361–376, Band 79, Teil II, S. 489–527, Band 80, Teil III, S. 437–490, Band 81, Teil VI, S. 109–139.

Secchi, Angelo: Analyse spectrale de la lumière de quelques étoiles, et nouvelles observations sur les taches solaires und Nouvelles recherches sur l'analyse spectrale de la lumière des étoiles. In: *Comptes Rendus Hebdomadaires des Séances de l'Académie des Sciences*, Tome **63** (1866), S. 364–368 und 621–628.

Vogel, Hermann Wilhelm: Über die neuen Wasserstofflinien und die Spectra der weissen Fixsterne. In: *Astronomische Nachrichten* **96** (1880), Nr. 21, S. 327–330.

Wien, Wilhelm: Ueber die Energievertheilung im Emissionsspectrum eines schwarzen Körpers. In: *Annalen der Physik und Chemie* **58** (1896), Nr. 294, S. 662–669.

Wolfschmidt, Gudrun: Kosmochemie – Chemische Elemente im Kosmos – Meteoriten, Sterne, Kosmologie. In: Wolfschmidt, Gudrun (Hg.): *Kosmochemie – Geschichte der Entdeckung und Erforschung der chemischen Elemente im Kosmos.* Hamburg: tredition (Nuncius Hamburgensis; Band 50) 2022, S. 66–135, Kapitel 2.4.3: Auf dem Weg zur Klassifikation von Sternspektren, S. 93–97.

Wollaston, William Hyde: A Method of Examining Refractive and Dispersive Powers, by Prismatic Reflection. In: *Philosophical Transactions of the Royal Society of London* **92** (1802), p. 365–380, On the Dispersion of Light, S. 372 f.

Abbildung 3.1:
Heinrich Kayser (1853–1940)
(© royalsocietypublishing.org)

Heinrich Kaysers „*Handbuch der Spectroscopie*“ (1900–1934) und dessen Bedeutung für die Astrophysik

Xian Wu (Dresden)

Abstract: Heinrich Kaysers „*Handbuch der Spectroscopie*"' (1900–1934) and its Significance for Astrophysics

The *Handbuch der Spectroscopie* (1900–1934) in eight volumes, mainly written by Heinrich Kayser (1853–1940), had a worldwide good reputation among scientists who were dealing with spectroscopy. The volumes were reviewed by contemporaries including astrophysicists in distinguished journals. As a standard reference work in the area of spectroscopy, the handbook made significant contributions to the development of astrophysics in the 20th century.

Zusammenfassung

Das hauptsächlich von Heinrich Kayser (1853–1940) verfasste achtbändige *Handbuch der Spectroscopie* (1900–1934) genoss weltweit eine hohe Reputation unter Wissenschaftlern, die sich mehr oder weniger mit spektroskopischen Fragen beschäftigten. Die Bände wurden von Zeitgenossen, darunter auch Astrophysikern, in namhaften Zeitschriften rezensiert. Als ein Standard-Nachschlagewerk auf dem Gebiet der Spektroskopie leistete das Handbuch wichtige Beiträge zur Entwicklung der Astrophysik im 20. Jahrhundert.

3.1 Einleitung

1860 veröffentlichten Gustav Kirchhoff (1824–1887) und Robert Bunsen (1811–1899) in *Annalen der Physik und Chemie* einen Beitrag mit dem Titel *„Chemische Analyse durch Spectralbeobachtungen“*, wobei eine bahnbrechende Methode (entdeckt bereits 1859) beschrieben wurde, mit der die chemische Zusammensetzung eines Stoffs allein durch die Analyse der Spektrallinien identifiziert werden konnte [Kirchhoff 1860]. Die Forscher schrieben zum Schluss ihres Beitrags:

> *„Wir haben uns in dieser Abhandlung darauf beschränkt, die Spectren der Metalle der Alkalien und alkalischen Erden und diese auch nur in so weit zu untersuchen, als es für die Analyse irdischer Stoffe nöthig ist. Wir behalten uns vor, diesen Untersuchungen die weitere Ausdehnung zu geben, die wünschenswerth ist in Beziehung auf die Analyse irdischer Körper und auf die Analyse der Atmosphären der Gestirne.“*

Diese Entdeckung aus Heidelberg wird oftmals als die Geburtsstunde der Astrophysik angesehen, obwohl der Fachbegriff *„Astrophysik“* erst 1865 von Johann Karl Friedrich Zöllner (1834–1882) in Leipzig eingeführt wurde.

Nach ihrer Entdeckung entwickelte sich die Spektralanalyse schnell, vor allem in den Disziplinen Chemie und Astronomie [Wolfschmidt (2023)]. Im Vergleich dazu blieb die physikalische Grundlage der Spektroskopie jedoch weit hinter. Obwohl die erste Serienformel für Spektrallinien des Wasserstoffs bereits 1885 von dem Schweizer Johann Jakob Balmer (1825–1898) entdeckt wurde, stand die tiefe physikalische Begründung noch offen. Dies war ohne die Kenntnisse des Atomaufbaus nicht möglich. Die Entdeckung des Elektrons erfolgte gegen Ende des 19. Jahrhunderts, während das Bohrsche Atommodell erst im Jahre 1913 entworfen wurde.

Nichtsdestotrotz verwendeten Astronomen die mächtige Technik der Spektroskopie für die Untersuchung von den chemischen Zusammensetzungen der Himmelskörper und erschlossen damit völlig neue Forschungsgebiete der Himmelskunde. Edwin Dunkin (1821–1898) in Greenwich definierte 1869 Astrophysiker als diejenigen, die spektroskopische Beobachtungen der Sonne und der Sterne ausüben und die Spektren von diesen mit den im Labor erzeugten Spektren von Metallen und Gasen vergleichen [DeVorkin 1997]. 1895 wurde in den USA das *Astrophysical Journal* als die weltweit erste Zeitschrift für Astrophysik gegründet. Da die Spektroskopie mittlerweile eine besonders wichtige Rolle in der Astrophysik spielte, fungierte diese neue Zeitschrift für mehr als zehn Jahre als ein Forum der spektroskopischen Daten [DeVorkin (1997)].

Lehrbuch

der

Spektralanalyse

von

Dr. Heinrich Kayser,

Privatdocent an der Universität zu Berlin und Assistent am Physikalischen Institut.

Mit 87 in den Text gedruckten Holzschnitten und 9 lithogr. Tafeln.

Springer-Verlag Berlin Heidelberg GmbH

1883.

Abbildung 3.2:
Heinrich Kayser: *Lehrbuch der Spektralanalyse* (1883)

Bis zu Ende des 19. Jahrhunderts erschienen zwar bereits zahlreiche Bücher über Spektroskopie bzw. Spektralanalyse in z. B. deutscher, englischer und französischer Sprache. Es fehlte allerdings ein Werk, das eine möglichst vollständige Übersicht über die bisherigen Kenntnisse der Spektroskopie zusammenfasste, und zwar mit Angaben von möglichst vollständigen Literaturen sowie kritischen Kommentaren dazu. Ein solches Nachschlagwerk sollte die Bedürfnisse der Forscher und Anwender der schnell anwachsenden Spektroskopie befriedigen. Selbstverständlich könnte die Verfassung des Werks außerordentlich schwierig sein, da die Sammlung, Analyse und Zusammenstellung der in verschiedenen Zeitschriften und Büchern zerstreuten spektroskopischen Literaturen besonders zeitaufwendig sind. Heinrich Kayser (1853–1940), damals Professor für Physik an der Universität Bonn, übernahm diese Aufgabe. Er publizierte, zuweilen mit Unterstützungen seiner Zeitgenossen, von 1900 bis 1934 das achtbändige *Handbuch der Spectroscopie.* Dieses großartige Werk war zwar auf das gesamte Gebiet der Spektroskopie gerichtet, es erlangte jedoch großen Ruhm unter zeitgenössischen Astrophysikern.

Im Folgenden werden, nach einer kurzen Vorstellung zur Person Heinrich Kayser, die 8 Bände des Handbuchs angesprochen. Der Verfasser dieses Beitrags konnte alle Bände aus dem Magazin der Sächsischen Landesbibliothek – Staats- und Universitätsbibliothek (SLUB) in Dresden bereitstellen lassen und näher betrachten. Zudem wurden die meisten in bedeutenden Fachzeitschriften erschienenen Rezensionen herausgefunden. So konnten die Beurteilungen der zeitgenössischen Rezensenten evaluiert werden.

3.2 Heinrich Kayser – der Meister der klassischen Spektroskopie

Heinrich Kayser (Vollname: Johannes Gustav Heinrich Kayser, Abb. 3.1) wurde am 10.03.1853 in Bingen am Rhein geboren. Kurz nach seiner Geburt zog die Familie zuerst nach Wiesbaden, dann nach Halle um. In Halle begann Kaysers Schulbildung, die sich später in Berlin fortsetzte. 1873 begann Kayser Physik an der Universität Straßburg zu studieren. Dort lernte er praktische Laborarbeiten bei dem namhaften Experimentalphysiker August Kundt (1839–1894). Kundts große Wertschätzung für Experimente gegenüber Mathematik für die Physikforschung beeinflusste den jungen Kayser in hohem Maß. 1877 veröffentlichte er in *Annalen der Physik und Chemie* seinen ersten wissenschaftlichen Beitrag über die Bestimmung von thermodynamischen Konstanten durch akustische Experimente [Kayser (1877)].

Abbildung 3.3:
Carl Runge (1856–1927)

(University of Chicago Photographic Archive, Special Collections Research [apf6-04312], Hanna Holborn Gray Special Collections Research Center, University of Chicago Library)

Kayser promovierte 1879 bei Hermann Helmholtz (1821–1894) mit dem Thema *„Ueber den Einfluss der Intensität des Schalles auf seine Fortpflanzungsgeschwindigkeit“* [Kayser 1879] und habilitierte sich anschließend. In Berlin setzte er einerseits die Forschung experimenteller Akustik fort. Andererseits begann er, sich mit dem Phänomen der Adsorption, der wissenschaftlichen Photographie sowie der Spektroskopie zu beschäftigen. Die Technik der Photographie lernte Kayser bei Hermann Wilhelm Vogel (1834–1898), einem Chemiker an der Technischen Hochschule Berlin-Charlottenburg.

Kaysers Interesse an der Spektroskopie wurde erregt durch die Experimentalvorlesungen von seinem Kollegen Ernst Hagen (1851–1923), der einst bei Kirchhoff und Bunsen studiert hatte. Immer mehr Zeit investierte er in diesem damals für Physiker eher mysteriösen Bereich. So publizierte er bereits 1883 das *Lehrbuch der Spektralanalyse* mit 17 Kapiteln bzw. ca. 380 Seiten [Kayser (1883), Abb. 3.2]. Das Buch enthält 3 Teile, nämlich *„Die Emission des Lichtes“*, *„Die Absorption des Lichtes“* sowie *„Die Spektren der chemischen Elemente und ihrer wichtigsten Verbindungen“*. Im Vorwort erklärte der Autor die zwei Ziele des Werks: Einerseits möchte er den Studierenden eine vollständige Übersicht über die Gesetze, Methoden und Apparate der Spektralanalyse in kurzer Form ermitteln. Andererseits sollte das Buch den möglichst vollständigen Forschungsstand des Gebiets darstellen, was für die Forschenden nützlich sein könnte. Im gleichen Jahr bot Kayser an der Universität Berlin eine Vorlesung zur Spektralanalyse an, die zahlreiche Studierende anzogen. Um spektroskopische Forschungen selbst auszuüben, erhielt Kayser durch Vermittlung von Helmholtz ein Konkavgitter von dem amerikanischen Physiker Henry Rowland (1848–1901).

1885 wurde Kayser als Professor für Physik an die Technische Hochschule Hannover berufen. Dort arbeitete er mit dem sich ebenfalls für Spektroskopie interessierenden Mathematiker Carl Runge (1856–1927, Abb. 3.3) zusammen. Ihr Ziel war, effektive mathematische Formeln für die Spektrallinien chemischer Elemente zu entwickeln, wobei Kayser sich in der präzisen Messung der Spektrallinien konzentrierte, während Runge anhand dieser Daten die sich dahinter verborgenen mathematischen Regeln herausfinden sollte. Die Kooperation führte zu zehn Forschungsartikeln innerhalb von sechs Jahren, wobei Formeln zur Berechnung von Spektrallinien für Elemente der Hauptgruppen 1 bis 5 sowie der Nebengruppen 1 und 2 entwickelt wurden. Diese Formeln wurden jedoch von der Arbeit des schwedischen Physikers Johannes Rydberg (1854–1919) überholt, da die von ihm entwickelte Formel sich als einfacher und allgemeingültiger erwies.

1894 veröffentlichte Kayser zwei Artikeln über die Spektren von Kometen in der Zeitschrift *Astronomische Nachrichten*. So wies er darauf hin, dass Kome-

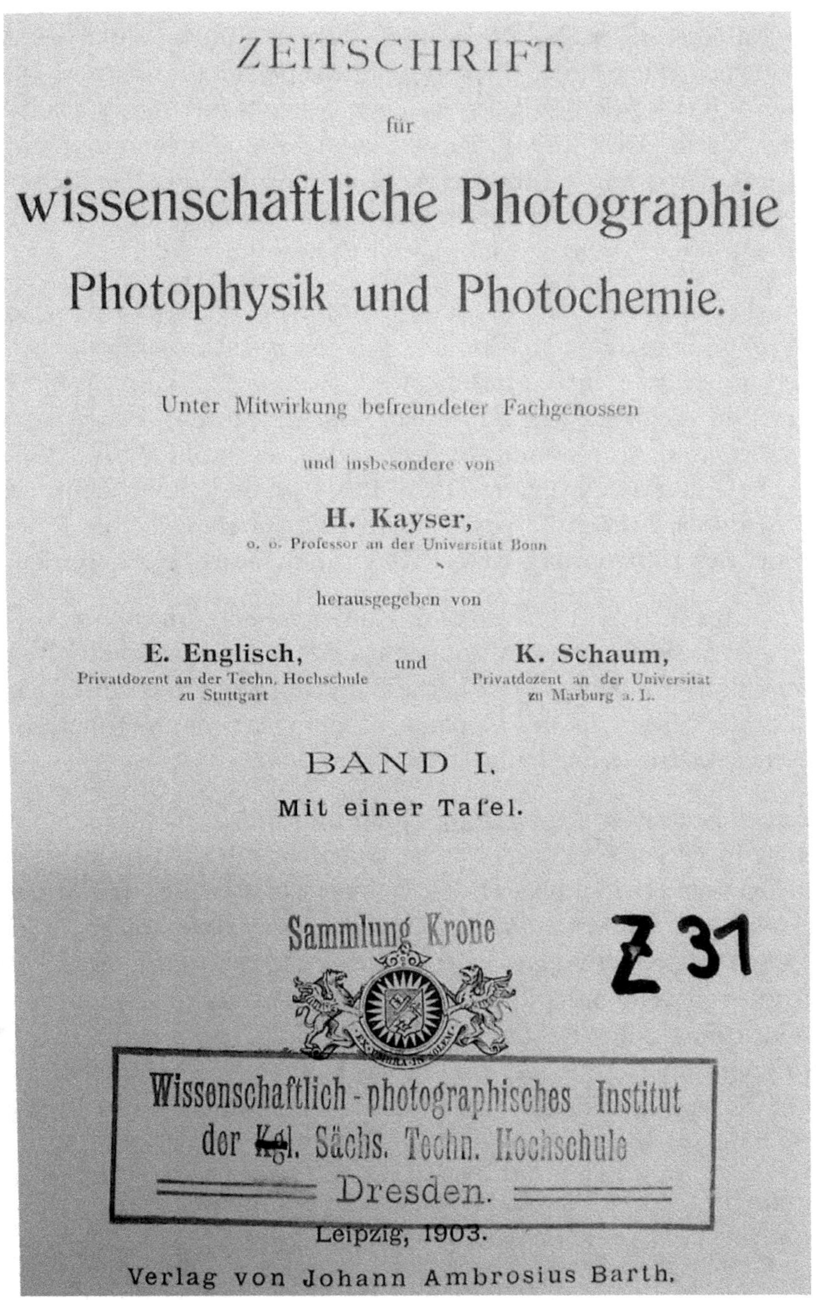

ZEITSCHRIFT

für

wissenschaftliche Photographie

Photophysik und Photochemie.

Unter Mitwirkung befreundeter Fachgenossen

und insbesondere von

H. Kayser,

o. ö. Professor an der Universität Bonn

herausgegeben von

E. Englisch, Privatdozent an der Techn. Hochschule zu Stuttgart

und

K. Schaum, Privatdozent an der Universität zu Marburg a. L.

BAND I.

Mit einer Tafel.

Leipzig, 1903.

Verlag von Johann Ambrosius Barth.

Abbildung 3.4:
Titelseite des ersten Hefts der *Zeitschrift für wissenschaftliche Photographie, Photophysik und Photochemie* (Leipzig: Johann Ambrosius Barth 1903)

ten Kohlenwasserstoffe enthalten könnten [Kayser (1894a)] und das Aussehen der Kometenspektren von der Spaltweite des verwendeten Instruments abhängig sein sollte [Kayser (1894b)]. Im gleichen Jahr wurde Kayser als Nachfolger von Heinrich Hertz (1857–1894), der am Anfang des Jahres im erstaunlich jungem Alter gestorben war, an die Universität Bonn berufen. Die Schwerpunkte seiner wissenschaftlichen Arbeiten in Bonn lagen in Spektren von Sternen und Elementen sowie der Standardisierung der Spektralmessung.

1903 wurde die *Zeitschrift für wissenschaftliche Photographie, Photophysik und Photochemie* gegründet (Abb. 3.4). Als der wichtigster fachlicher Mitwirkender betonte Kayser im Gründungswort [Kayser (1903)]: *„Jeder zugefügte Stein trägt zum Gelingen des Baues bei, jede Arbeit muß dankbar anerkannt werden, vielleicht erweist gerade sie sich einmal von größter Bedeutung.“* Diese Aussage trifft Ejnar Hertzsprung (1873–1967) besonders zu, denn seine 1905 und 1907 in dieser Zeitschrift veröffentlichten zwei Beiträge als Pionierarbeit des Konzepts des Hertzsprung-Russell-Diagramms angesehen werden.

Vom 31.07. bis 05.08.1913 fand die 5. Konferenz der *International Union for Cooperation in Solar Research* in Bonn statt. Als Gastgeber konnte Kayser über 100 Wissenschaftler aus verschiedenen Ländern in Bonn begrüßen. Kurz davor wurde ein neues physikalisches Gebäude eingerichtet, das seitdem als Sitz des Physikalischen Instituts der Universität Bonn dient.

Nach seiner Emeritierung im Jahre 1920 publizierte Kayser noch zwei weitere Bücher, in denen spektroskopische Daten ausführlich in Tabellen aufgelistet sind [Kayser (1925), Kayser (1926)]. Sein letzter wissenschaftlicher Beitrag erschien 1938 mit dem Titel *„Statistik der Spektroskopie“* in *Physikalische Zeitschrift*, wobei er einen Überblick über die weltweiten Forschungen zur Spektroskopie seit 1800 anhand der Analyse der Anzahl der Veröffentlichungen nach Ländern darstellte [Kayser (1938)]. Ein Jahr davor fertigte er das Manuskript *Erinnerungen aus meinem Leben* in fünf Exemplaren, wovon heute nur noch ein Exemplar existiert. 1996 wurde das Manuskript in Buchform von Wissenschaftshistorikern herausgegeben [Dörries & Hentschel (1996)].

Kayser starb am 14.10.1940 in Bonn. Arnold Sommerfeld (1868–1951) nannte ihn *„Nestor der deutschen Physik“* [Sommerfeld (1941)].

Kayser wurde dreimal (1905, 1916 und 1917) zum Nobelpreis nominiert. Die Einheit der Wellenzahl cm^{-1} des CGS-Systems wurde nach ihm benannt. Heute findet man den Namen von ihm (und seinen Nachfolgern) auf einer Gedenktafel neben der Eingangstür des Physikalischen Instituts der Universität Bonn (Nußallee 12, Abb. 3.5).

Abbildung 3.5:
Das heutige Physikalische Institut der Universität Bonn
und die Gedenktafel für Kayser neben dem Eingang

(Fotos: Xian Wu)

3.3 Kaysers Meisterwerk – *Handbuch der Spectroscopie*

3.3.1 Vorbereitungen

Schon in der Berliner Zeit begann Kayser spektroskopische Literatur zu sammeln. Dieser Fleiß setzte sich in Hannover und Bonn fort. Seine umfangreichen Kenntnisse in der Spektroskopie wollte er nutzbringend verwenden. Für ihn wäre dies durch zwei Wege realisierbar: entweder experimentelle Studien zur Ergänzung der Wissenslücke in der Spektroskopie oder eine kritische literarische Zusammenstellung der bisherigen spektroskopischen Kenntnisse. Da am Anfang seiner Bonner Zeit die experimentelle Ausstattung am dortigen Physikalischen Institut sehr schlicht war, entschied sich Kayser für den zweiten Weg, nämlich die Verfassung eines Handbuchs für alle bisherigen Kenntnisse der Spektroskopie.

Für den künftigen astrophysikalischen Teil des Handbuchs fühlte sich Kayser jedoch nicht sicher. Er kannte zwar astrophysikalische Instrumente aus fachlichen Texten, hatte diese aber nie mit eigenen Augen gesehen. Um diese Erfahrung nachzuholen, beschloss er wichtige astrophysikalische Observatorien zu besuchen. Zu seiner Zeit war in Deutschland das Potsdamer Observatorium die einzige bedeutende Einrichtung solcher Art. Allerdings wollte Kayser dort nicht hingehen. Der Grund lag darin, dass damals zwischen ihm und dem Direktor des Potsdamer Observatoriums Hermann Carl Vogel (1841–1907) eine ungünstige Beziehung bestand. H. C. Vogel kritisierte zuvor unfreundlicherweise Kaysers *Lehrbuch der Spektralanalyse.* Der namhafte Astronom war besonders unzufrieden, dass Kayser in diesem Buch die Photographie als eine sehr wichtige Methode für die Weiterentwicklung der spektroskopischen Messung erachtete. Zudem schätzte H. C. Vogel die Bedeutungen der in *Astronomische Nachrichten* veröffentlichten zwei Beiträge von Kayser über Kometenspektren gering. Wie Kayser in seiner Autobiographie erzählte, verschwand der Groll erst im Jahre 1903, als H. C. Vogel schriftlich um Kaysers Hilfe bat [Kayser (1996)].

Kayser blickte daher nach den USA. Dort standen ja drei berühmte Sternwarten für die astrophysikalische Forschung zur Verfügung, nämlich das Lick Observatory, das Harvard College Observatory sowie das Yerkes Observatory. Seine Reise in Amerika war fruchtbar, da er nicht nur wichtige Forschungseinrichtungen besucht, sondern auch bedeutende amerikanische Physiker und Astronomen persönlich kennengelernt und mit diesen befreundet hatte.

3.3.2 *Handbuch der Spectroscopie* Band I

Der erste Band des Handbuchs wurde im Jahre 1900 beim Verlag von S. Hirzel in Leipzig publiziert [Kayser (1900), Abb. 3.6, S. 114]. Der Verlag übernahm auch die Publikationen der folgenden Bände des Handbuchs. Band I enthält 6 Kapiteln bzw. 781 Seiten.

Im Vorwort schrieb Kayser:

> „*Eine Hauptbedingung für ein derartiges Werk ist möglichste Vollständigkeit, und ich habe keine Mühe gescheut, um mich ihr zu nähern. Die Auffindung der Litteratur ist auf wenigen anderen Gebieten so schwierig, wie auf unserem: in physicalischen, chemischen und astronomischen Zeitschriften wird man naturgemäss suchen, aber es sind auch in mathematischen, medicinischen, botanischen, photographischen, electrotechnischen und anderen technischen Zeitschriften Arbeiten zerstreut, weil überall das Spectroscop seine Anwendung findet* [...] *Ich habe beim Sammeln der Litteratur folgendes Verfahren eingeschlagen: etwa 40 der wichtigsten Zeit- und Gesellschaftsschriften wurden vom Jahre 1860 an genau durchgesehen, natürlich nicht nur die Titel, sondern der Inhalt. Alle spectroscopischen Citate wurden notirt und ihnen nachgegangen, u.s.w.* [...] *Wenn es mir so auch gelungen ist, etwa 7000 Abhandlungen oder Notizen zusammenzubringen, und wenn ich auch glaube, dass mir keine Arbeit von grosser Bedeutung entgangen ist, so bin ich doch überzeugt, dass ich noch Vieles übersehen habe. Vollständigkeit zu erreichen ist für einen Menschen unmöglich.*“
>
> Weiter schrieb er:
>
> „*Für ebenso wichtig, wie Vollständigkeit, halte ich eine kritische Zusammenfassung der Ansichten.* [...] *Ich habe daher überall, wo ich mir eine Meinung gebildet habe, und wo ich Methoden und Schlüsse für falsch oder auch nur für unbewiesen halte, frei Kritik geübt.*“

Darüber hinaus erläuterte Kayser seinen Plan für zukünftige Bände des Handbuchs, wobei der fünfte Band „*die Spectroscopie der Himmelskörper zur Darstellung bringen soll.*“

Dem Vorwort folgt eine Einleitung, in der einige Bemerkungen zur Benutzung des Handbuchs wie z. B. die Abkürzungen der zitierten Zeitschriften, die allgemeinen Werke über Spektralanalyse, die Angaben der Einheiten und der physikalischen Parameter, die Darstellungsmethoden der Spektren usw. erklärt werden. Danach beginnt das erste Kapitel im Umfang von 128 Seiten über die „*Geschichte der Spectroscopie*“. Für Kayser fängt die Geschichte dieses Gebiets

HANDBUCH

DER

SPECTROSCOPIE

VON

H. KAYSER

PROFESSOR AN DER UNIVERSITÄT BONN.

ERSTER BAND.

LEIPZIG

VERLAG VON S. HIRZEL

1900.

Abbildung 3.6:
Titelseite des ersten Bands des *Handbuchs der Spectroscopie*
von Heinrich Kayser (Leipzig 1900).

(Sammlung der Sächsischen Landesbibliothek – Staats- und Universitätsbibliothek)

mit Isaac Newtons (1642–1726 jul./1643–1727 greg.) Untersuchungen von Licht an. Das Kapitel beschreibt wichtige Entdeckungen und Entwicklungen in der Spektroskopie beginnend von Newton entlang der Zeitachse bis zu Ende des 19. Jahrhunderts.

Kapitel 2 behandelt „*Die Erzeugung leuchtender Dämpfe*“. Dabei werden die drei verschiedenen Lichterzeugungsverfahren durch Flammen, Lichtbögen und Entladungen ausführlich beschrieben. Bei Kapitel 3 (geschrieben von Kaysers Assistent Heinrich Konen (1874–1948), der 1897 bei Kayser promovierte) und Kapitel 4 geht es jeweils um „*Prismen*“ und „*Diffractionsgitter*“ (die Theorie der Konkavgitter übernahm Runge) – die zwei wichtigen Gruppen der optischen Elemente zur Zerlegung des Lichts. Kapitel 5 beschreibt „*Die spectroscopischen Apparate*“, einschließlich der Photographie sowie der Fluoreszenz, der Phosphoreszenz und der Wärmestrahlung. Zum Schluss werden „*Die spectroscopischen Messungen*“ in Kapitel 6 erläutert. Kapitel 2 bis 6 sind direkt relevant zur Anwendung der Spektroskopie in der Astronomie. Aber auch Kapitel 1 enthält zahlreiche Informationen zur Geschichte und Bedeutung der Spektroskopie in der Astronomie.

Die Erscheinung des Bandes erregte zahlreiche Kommentare. So teilte der Verlag Kayser mit [Kayser (1996)]:

> „*Die Besprechungen sind, wohl ohne Ausnahme, so vorzüglich gewesen, wie sie selten ein Werk aufzuweisen hat; und diese rühmenden teilweise begeisterten Stimmen aus dem In- und Auslande werden Ihre Freude an der weiteren Arbeit wesentlich erhöht haben.*“

1901 erschien eine Rezension in der Zeitschrift Nature von dem deutsch-britischen Wissenschaftler Arthur Schuster (1851–1934), der bei Kirchhoff in Heidelberg promovierte und ab 1888 Professor für Physik in Manchester fungierte. Schuster begann seine Rezension mit dem folgenden Text:

> „*There are comparatively few men of science who can accurately handle a spectroscope and interpret its indications with assurance. The number of chemists, for instance, who could look at the spectrum of a Geissler tube, and pick out at once the lines of hydrogen, oxygen, nitrogen or carbon, is probably very small. No one denies the importance of the spectroscopic method, but its practice requires so long an apprenticeship and so severe a training, while the experimental facts are so numerous and the pit-falls so plentiful, that the physicists and chemists are inclined to shirk the whole subject and to leave it to the few who happen to have been brought up in a spectroscopic atmosphere.*“

Und er schrieb weiter:
„*Part of the cause of this apparent neglect is due to the want of a proper guide to lead the willing but bewildered student through the intricacies of a most diffuse and uninviting literature. We possess only a few short textbooks which are quite insufficient for any serious requirements, and various catalogues of papers relating to spectrum analysis which have proved absolutely useless. Prof. H. Kayser, well known as an authority on the subject, has undertaken what must prove to be the work of a lifetime. The first volume of his ‚Treatise of Spectroscopy' is now completed and will be welcomed by all who desire to know, as well as by those who already know, something of this branch of science.*"
Er meinte:
„*There can be only one opinion on the admirable manner in which Prof. Kayser has accomplished his task. He has succeeded in giving a clear and complete account of his subject, and at the same time avoided overburdening his book with details, which the reader can always find in the original papers, to which complete references are given.*"

Der Rezensent kommentierte den Inhalt des Bandes mit kritischen Anmerkungen zu einigen speziellen Themen bezüglich der spektroskopischen Instrumente und Theorie. Zum Schluss lobte er das erfolgreiche Werk und dessen Verfasser nochmals und er hoffte zugleich die baldige Erscheinung des zweiten Bandes.

Eine weitere Rezension von dem amerikanischen Astrophysiker Henry Crew (1859–1953) erschien 1902 in *Astrophysical Journal* [Crew (1902)]. Crew promovierte 1887 bei Rowland auf dem Gebiet der Astrospektroskopie an der Johns Hopkins University, forschte von 1891 bis 1892 am Lick Observatory und war von 1892 bis 1933 Professor für Physik an der Northwestern University. Alle Kapiteln des Bandes wurden von Crew kommentiert und hoch geschätzt. Er schrieb zum Schluss in der Rezension:

„*Concerning the volume as a whole, a reader gets the impression that the author has gone through the entire periodical literature of physical science with a drag net from which nothing has escaped. The result is that these pages contain some matters which will probably prove new to the most accomplished spectroscopists of the world. In short, the work is of such a character that its possession is not only desirable but indispensable to every serious student of the science.*"

Die Erscheinung des ersten Bandes des Handbuchs war somit ein großer Erfolg. Kayser war besonders froh, dass Lord Rayleigh (1842–1919) dem Werk sehr hohe Wertschätzung gab, indem er es das zweitbeste Buch für ihn nannte, nur hinter dem *Handbuch der Physiologischen Optik* von Helmholtz [Kayser (1996)].

3.3.3 *Handbuch der Spectroscopie* Band II

Der zweite Band erschien im Jahre 1902 [Kayser (1902)]. Im Vorwort schrieb Kayser:

> *„In Bezug auf Litteraturangaben habe ich wieder gesucht Vollständigkeit zu erreichen.“*

Der Band enthält 9 Kapitel. Kapitel 1 behandelt *„Emission und Absorption“*, wobei im Wesentlichen das Kirchhoffsche Strahlungsgesetz diskutiert wird. Kapitel 2 und Kapitel 3 behandeln jeweils *„Strahlung fester Körper“* und *„Strahlung der Gase“*. Bei Kapitel 4 geht es um *„Verbindungsspectra. Mehrfache Spectra“*, gefolgt von Kapitel 5 über den *„Einfluss von Druck, Temperatur, Entladungsart auf die Spectra“* sowie Kapitel 6 über *„Das Aussehen der Spectrallinien“*. *„Das Dopplersche Princip“* ist das Thema von Kapitel 7, das von Konen geschrieben wurde. Kapitel 8 behandelt *„Gesetzmässigkeiten in den Spectren“*, wobei physikalische Gesetze zu Bandenspektren und Linienspektren ausführlich diskutiert werden. Das letzte Kapitel beschreibt *„Schwingungen des Lichtes im magnetischen Felde“*, wobei Runge dort die Inhalte über Zeeman-Effekt übernahm. Alle Kapitel konzentrieren sich auf den physikalischen Aspekten der Spektroskopie und sind hochrelevant zur Astrophysik.

Zwei wichtige Spektroskopiker bzw. Astrophysiker hatten diesen Band rezensiert. Einer war wieder Schuster. Seine Rezension erschien 1903 in *Nature* [Schuster (1903)]. Sie beginnt mit dem Satz: *„The second volume of this important work follows the first after a remarkably short interval of time“*, gibt Aufschluss über jedes Kapitel und fasst am Ende folgendes zusammen:

> *„Prof. Kayser may congratulate himself on the successful completion of this volume, which is full of suggestive criticism. Its value is enhanced by the fact that it brings the gaps in our knowledge prominently before us. Anyone wishing to advance by original research a science which is destined to clear up the secrets of molecular and atomic constitution will find Prof. Kayser's work full of promising starting points.“*

Der andere war der amerikanische Astrophysiker Georg Ellery Hale (1868–1938). Hale studierte Physik am Massachusetts Institute of Technology und wurde 1892 Professor für Astronomie an der University of Chicago. Er gründete jeweils 1895, 1897 und 1904 das *Astrophysical Journal*, das Yerkes Observatory und das Mount Wilson Observatory. Auch die Gründung des späteren Palomar Observatory ging auf seine Bemühung zurück. Die Rezension erschien 1903 in *Astrophysical Journal* mit großer Wertschätzung des Bandes [Hale (1903)]. Er schrieb:

> „*The first volume of the Handbuch, with its admirable description and critical account of spectroscopic instruments, has doubtless been a chief book of reference to all spectroscopists since the time of its appearance. The second volume deals with a wide range of subjects, and is even more fascinating than the first.*"
> Zu dem zweiten Kapitel kommentierte er:
> „*To the astrophysicist the summary in the second chapter of the remarkable recent advances, both experimental and theoretical, in our knowledge of the radiation of solid bodies will be of special importance in connection with the determination of the Sun's temperature.* [...] *The work described in this chapter, and also, indeed, in every other chapter of the book, affords striking illustration of the dependence of the astrophysicist upon the experimental and theoretical investigations of the physicist, with whose work he cannot be too familiar.*"

Zum Schluss der Rezension schrieb Hale:

> „*Even from this brief mention of some of the principal contents of the volume, it will be seen that the work of every spectroscopist must be greatly facilitated by having the book constantly at hand. The suggestions as to problems which still require investigation are most valuable, and may be especially commended to those who are seeking subjects for research. [...].*"

3.3.4 *Handbuch der Spectroscopie* Band III und Band IV

Band III und Band IV erschienen jeweils 1905 und 1908 [Kayser (1905), Kayser (1908)]. Sie behandeln die Themen Absorption, Dispersion, Phosphoreszenz und Fluoreszenz. Band III enthält die Kapitel „*Apparate und Methoden zur Untersuchung der Absorption*", „*Die Veränderlichkeit der Absorptionsspectra*", „*Beziehungen zwischen Absorption und Constitution organischer Körper*" (vom britischen Chemiker Walter Noel Hartley (1846–1913) auf Englisch geschrieben und Kayser ins Deutsche übersetzt), *„Absorption ausgewählter Stoffe“* sowie *„Alphabetisches Verzeichniss der bis jetzt bekannten Absorptionsspectra“*. Band IV enthält die Kapitel *„Natürliche Farbstoffe der Pflanzen“*, *„Die Farbstoffe von Blut, Harn, Galle“*, *„Thierische Farbstoffe“*, *„Dispersion“* (geschrieben von dem Physiologen Eduard Pflüger (1829–1910), *„Phosphorescenz“* und *„Fluorescenz“* (letzterer verfasst von Konen). Eine Gemeinsamkeit dieser zwei Bände besteht darin, dass sie im Vergleich zu den ersten zwei Bänden großes Gewicht auf die chemischen Aspekte der Spektroskopie liegen.

Im Vorwort von Band III schrieb Kayser:

> *„Die Arbeit an diesen beiden Bänden ist, wie der kundige Leser leicht sehen wird, eine äusserst mühevolle und doch wenig angenehme gewesen; es ist Handlangerarbeit, die sich meist auf Sammeln und Sortiren eines minderwerthigen Beobachtungsmaterials beschränkt. [...] wenn auch theoretisch wenig Ergebnisse gewonnen sind, so ist doch für practische Zwecke eine Sammlung des vorliegenden Beobachtungsmateriales wichtig, und eine solche Sammlung ist auch die Vorbedingung für weiteres theoretisches Arbeiten.“*

1905 erschienen zwei englische Rezensionen zu Band III. Der amerikanische Physiker und Chemiker William Weber Coblentz (1873–1962) schrieb im *Astrophysical Journal* [Coblentz (1905)]:

> *„The frequent references to Kayser's Spectroscopie in recent papers on spectroscopic subjects show that it is filling a long-felt need. [...] this volume fulfils the promise of the earlier ones in completeness and thoroughness of treatment.“*

Für ihn war der Band *„a standard of reference“*. Coblentz promovierte an der Cornell University in Physik und war von 1903 bis 1905 Forscher an der Carnegie Institution of Washington und von 1905 bis 1945 Forscher am *National Bureau of Standards*. Er war eine Schlüsselperson für die Frühentwicklung der Infrarot-Spektroskopie in der organischen Chemie.

Die in *Nature* erschienene Rezension wurde von dem Chemiker Edward Charles Baly (1871–1948) verfasst [Baly (1905)]. Als Dozent für Spektroskopie an der University of London fand Baly diesen Band hochinteressant. Trotz seiner kritischen Anmerkungen zu manchen Ansichten der Absorption, fasste er folgendes zusammen:

> *„Of the great value of this book it is impossible to speak too highly; it is sufficient to say that it will rank as the standard work upon*

> *absorption. All who read it will appreciate to the full the great care Prof. Kayser has bestowed upon it and the immense labour involved in dealing with the mass of literature upon the subject.“*

Zu Band IV sind wieder zwei Rezensionen, erschienen jeweils im *Astrophysical Journal* und in *Nature*, zu finden. Die Rezension im *Astrophysical Journal* wurde von dem Physiker und Astronom Robert Williams Wood (1868–1955), damals Professor für Experimentalphysik an der Johns Hopkins University verfasst [Wood (1908)]. Er stellte eine sehr positive Beurteilung bezüglich der Autorität und des Umfangs des Bandes dar, und machte das folgende Fazit:

> *„Upon the whole I am inclined to regard this book as the most interesting which has appeared within recent years, and feel tempted to insert it in the oft-quoted saying that ‚The student armed with the calculus and the spectroscope cannot fail to discover new and important laws of nature.‘ For my own part, if my library were to be reduced to two works, I think that I should select Kayser's Handbuch and Lord Rayleigh's Collected Papers.“*

Der Rezensent in *Nature* schätzte den Band ebenfalls hoch, obwohl der Name des Rezensenten dort nicht angegeben wurde [Anonym (1908)].

3.3.5 *Handbuch der Spectroscopie* Band V und Band VI

Die Verfassung der Bände V und VI war methodisch anders als die früheren Bände, denn ab Band V handelt es sich um die Zusammenstellungen von spektroskopischen Daten der Elemente sowie die Kommentare dazu. Wie Kayser selbst im Vorwort zu Band V erwähnte [Kayser 1910]:

> *„Während die ersten Bände hauptsächlich für solche bestimmt sind, die selbst in die Forschung eingreifen wollen, soll dieser und der folgende Band ein Nachschlagebuch für Jeden sein, der irgend welche spectroscopische Daten braucht. Solche Daten zu finden, war bei der bisher vorliegenden zusammenfassenden Litteratur außerordentlich zeitraubend, wenn nicht unmöglich* [...] *So, denke ich, werden diese Bände einerseits viel Zeit sparen, andererseits zu weiteren Arbeiten anregend wirken.“*

Band V erschien 1910 und enthält die spektroskopischen Daten zu den Elementen, deren Symbole alphabetisch bis zur Anfangsbuchstabe N zugeordnet sind, wobei die Teile zu Brom, Chlor, Cäsium, Jod, Kalium und Lithium von

Konen verfasst wurden. Die übrigen Elemente sind in Band VI, der im Jahre 1912 erschien, enthalten [Kayser (1912)]. Die Teile zu Natrium, Rubidium, Schwefel, Selen und Tellur wurden von Konen übernommen.

Im Vorwort dieses Bandes erläuterte Kayser den Grund, warum ein Band extra für die Astrophysik nicht verfasst werden konnte. Das Gebiet der Spektroskopie entwickelte sich nämlich zu schnell, so dass die Anzahl der Literatur enorm gewachsen war. Er schrieb:

> *„Wenn ich die mir vorliegende astrophysikalische Literatur überschaue, so ist mir klar, dass der dafür beabsichtigte eine Band nicht annähernd ausreicht; es würden wohl 3 Bände dafür erforderlich werden, und ich würde wohl eine Zeit von 8 Jahren dazu gebrauchen. Aber ich bin keinen Augenblick darüber in Zweifel, dass ich inzwischen zu alt geworden bin, um ein solches neues Werk zu unternehmen* [...] *So bin ich, freilich mit Bedauern, zu dem Entschluss gekommen, die Darstellung der mich ganz besonders interessirenden astrophysikalischen Anwendungen jüngeren Händen zu überlassen* [...].“

In der Rezension von Crew im *Astrophysical Journal* zu Band V rühmte der Rezensent [Cre 1911]:

> *„It is, in fact, the first orderly and comprehensive summary and discussion of results in a science which has already celebrated its jubilee.“*

Ihm war klar, dass die Spektroskopie gerade eine dramatische Entwicklung erlebte und viele früheren Standardmethoden mittlerweile nicht mehr gültig waren. Aber *„we stand on the shoulders of our predecessors“*. Auch der Verlag wurde gelobt:

> *„A publisher who will undertake the issue of six large volumes devoted to a science so utterly ‚useless‘ (from the point of view of ‚the man on the street‘) as spectroscopy is at once a patron of, and an honor to, pure science. The joint service of publisher and author is not likely to be over-estimated.“*

Baly rezensierte Band V in *Nature* und meinte [Baly (1911)]:

> *„There is no doubt that this volume is a very worthy follower of the first four in the series, and must prove an indispensable addition to the library of everyone interested in emission spectra.“*

Band VI bekam wieder eine Rezension von Crew im *Astrophysical Journal* [Crew (1913)]. Wie in seinen bisherigen Rezensionen wurde auch in dieser letzten Rezension von ihm die Leistung von Kayser anerkannt. Er schrieb:

> *„The stupendous undertaking upon which Professor Kayser entered more than a quarter of a century ago was the production of a historical compendium which should contain practically all that is known concerning the subject of spectroscopy. The completion of such a work is a noteworthy event in the annals of this science.“*
> Und
> *„if it were not already true that an easy reading knowledge of German is an absolute necessity for any serious student of spectroscopy, the completion of this handbook would certainly make it true; for this work is the one indispensable treatment of its subject. Professor Kayser surely deserves, and as surely has, the congratulations and the thanks of his fellow-workers and fellow-students in all parts of the world.“*

3.3.6 *Handbuch der Spectroscopie* Band VII und Band VIII

Während bisher alle 2 bis 3 Jahre ein neuer Band erschien, kam Band VII erst 12 Jahre später nach der Erscheinung von Band VI auf die Welt [Kayser 1924]. Zudem war der 1924 publizierte Band nur der erste Teil der insgesamt drei Lieferungen, wobei die zweite und dritte Lieferungen jeweils im Jahre 1930 und 1934 erschienen [Kayser (1930), Kayser (1934)].

Ungewöhnlicherweise erschien Band VIII im Jahre 1932 [Kayser (1932)], d. h. in der Zwischenzeit nach der zweiten Lieferung von Band VII und vor dessen dritter Lieferung. Kayser ging in den Ruhestand im Jahre 1920. Sein Nachfolger war Konen, der ab Band VII als Co-Autor des Handbuchs fungierte.

Band VII und Band VIII sollten die in Band V und Band VI zusammengestellten Spektren der Elemente aktualisieren, da mittlerweile viele neue Forschungsergebnisse auf diesen Gebieten entstanden. Kayser erzählte in seiner Autobiographie, wie die Bände VII und VIII zustande gekommen war. Nach dem Ersten Weltkrieg wurde Kayser von *„einem amerikanischen Kollegen“* gefragt, ob sein Handbuch fortgesetzt werden konnte, da die bisherigen Bände dem aktuellen Bedarf nicht mehr genügten. Besonders ermuntert von Theodore Lyman (1874–1954) und William Meggers (1888–1966), entschied sich Kayser trotz der Schwierigkeit, dass den Bibliotheken in Deutschland während des Ersten Weltkriegs ausländische Zeitschriften nicht zugänglich waren, weitere Bände zu schaffen.

Er lud Konen ein, diese nun gemeinsam herauszugeben, denn er brauchte einerseits einen Co-Autor, der die theoretischen Teile mit vielen mathematischen Hintergründen verfassen konnte, andererseits möchte er diese Gelegenheit nutzen, um Konens Namen in der akademischen Welt weiter zu verbreiten. Konen akzeptierte Kaysers Einladung. Die Kooperation war leider für Kayser nicht glücklich. Wegen Konens Verzögerung konnte der erste Teil von Band VII viel später erscheinen als geplant. Danach hörte Konen sogar auf, weiter zu bearbeiten. Ein paar Jahre später teilte Konen mit, dass er wieder am Handbuch arbeiten konnte. Nach der Erscheinung des zweiten Teils von Band VII entschied sich Konen, die restlichen Arbeiten auf mehrere junge Kollegen aufzuteilen. Kayser war mit dieser Vorgehensweise nicht zufrieden. Nach der Erscheinung des ersten Teil von Band VIII hörte Konen komplett auf, die Arbeit fortzusetzen.

Nichtsdestotrotz wurden Band VII und Band VIII vom Verlag S. Hirzel publiziert, auch wenn teilweise mit ungewöhnlicher Erscheinungsreihenfolge. Band VII Teil I umfasst Elemente, deren Symbole bis zur Anfangsbuchstabe F haben. Teil II enthält diejenigen von G bis zu I. Von J bis zu N berücksichtigt Teil III. Der erste und zugleich der einzige Teil von Band VIII, der die Inhalte von Band VII aktualisieren sollte, fängt wieder von Buchstabe A an, und endet mit der Buchstabe C (Cu).

Der damalige Direktor vom Yerkes Observatory Edwin Frost (1866–1935) rezensierte die erste Lieferung von Band VII in dem von ihm mitherausgegebenen *Astrophysical Journal* [Frost (1924)]. Er schrieb:

> „*We congratulate Professors Kayser and Konen that they have been able to add a record of the work of spectroscopists for another decade of the great encyclopedia of practical spectroscopy for which the world is greatly indebted to Professor Kayser.*“

Für die 1930 erschienene zweite Lieferung von Band VII war keine Rezension bekannt. Im Gegensatz dazu erhielt der zwei Jahre später veröffentlichte Band VIII mehrere Rezensionen. Zwei englische Rezensionen erschienen 1933 jeweils in *Nature* und *Journal of Physical Chemistry*. Der englische Spektroskopiker Raynor Johnson (1901–1987) schrieb in seiner kurzer Rezension [Johnson (1933)]:

> „*There must always be a place of value for a compilation of reliable data such as at the time of publication vols. 5 and 6 represented. Vol. 7 and, now, Vol. 8 [...] are an admirable attempt to bring up to date this aspect of the subject of spectroscopy.*“

Und Baly schrieb in seiner Rezension für das *Journal of Physical Chemistry* – einer Zeitschrift für das chemische Gebiet [Baly (1933)]:

> *„Whilst one may legitimately wonder what may possible be the state of knowledge of the spectra of the elements described in the final section of this volume, compared with that of those now dealt with, one cannot but hold up to admiration the devoted labors of Professor Kayser and Professor Konen and of their six collaborators, which have given to the scientific world a book so urgently needed.“*

Die 1934 im *Astrophysical Journal* erschienene Rezension des bedeutenden amerikanischen Astrophysikers William Morgan (1906–1994), der damals am Yerkes Observatory forschte, war zwar kurz, aber enthielt die entscheidende Beurteilung zur Bedeutung des Bandes für die Astrophysik [Morgan (1934)]: *„This work is of the first importance to the practical astrophysicist.“*

Die ein Jahr zuvor in der *Zeitschrift für angewandte Mathematik und Mechanik* erschienene deutsche Rezension ist wohl die erste Rezension zu dem Handbuch in einer namhaften wissenschaftlichen Zeitschrift überhaupt [Simson (1933)]. Die Rezensentin war Clara von Simson (1897–1983), eine damals an der Universität Berlin forschende Physikerin, die 1923 bei Max von Laue (1879–1960) in Experimentalphysik promoviert hatte. Ihre vollständige Rezension lautet:

> *„Das starke Anschwellen der spezielle spectroskopische Daten enthaltenden Literatur verstärkt das Bedürfnis nach einem ‚Nachschlagewerk, das in handlicher Weise die erdrückende Fülle der Einzelergebnisse sichert, ordnet und die Literatur zusammenfaßt.‘ Es ist daher sehr zu begrüßen, daß in der hier vorliegenden Lieferung, wieder mit den Elementen beginnend, deren chemisches Symbol mit den Buchstaben A bis C anfängt, die Literatur-Zusammenstellung und -Bearbeitung bis Anfang 1932 in altgewohnter zuverlässiger und übersichtlicher Weise fortgeführt wird.“*

In der gleichen Zeitschrift erschien 1934 eine deutsche Rezension zur dritten Lieferung von Band VII [Anonym 1934]. Der Name des Rezensenten wurde leider nicht angegeben. Die vollständige Rezension lautet:

> *„Mit der vorliegenden dritten Lieferung ist der 7. Band des Handbuchs der Spectroscopie von Kayser und Konen abgeschlossen. Der Gesamtband umfaßt 57 Elemente in alphabetischer Reihenfolge von Argon bis Niobium. Bei der ungeheuren Fülle der spektroskopischen Publikationen der letzten Jahre haben die Verfasser den Anschluß des behandelten Stoffes an die Jetztzeit durch ein Annäherungsverfahren zu erreichen versucht. Die erste Lieferung des 7. Bandes*

vom Jahre 1923 ist durch die inzwischen erschienene erste Lieferung des 8. Bandes auf den Stand des Jahres 1932 gebracht. Die zweite Lieferung dieses 8. Bandes ist noch für 1934 in Aussicht gestellt, so daß dann in den beiden Bänden 7 und 8 die genannten 57 Elemente so nahe wie möglich an die Gegenwart herangeführt sind. Auch in dem neuen Band ist die bekannte und bewährte Form beibehalten worden. Die spektroskopischen Daten sind so vollständig angeführt, daß gerade auf dieser restlosen Umfassung eines umfangreichen Teilgebiets der Physik die hohe Bedeutung des Handbuchs für die Forschung beruht.“

3.3.7 Fazit

Tabelle 3.1 fasst einige Informationen zu den 8 Bänden des Handbuchs zusammen. Das gesamte Werk enthält über 7300 Seiten.

Tabelle 3.1:
Informationen zu den 8 Bänden des Handbuchs

Band	Erscheinungsjahr	Seitenzahl	Bemerkung
I	1900	781	–
II	1902	696	–
III	1905	604	–
IV	1908	1248	–
V	1910	853	–
VI	1912	1067	–
VII, Teil 1	1924	499	Konen als Co-Autor
VII, Teil 2	1930	251	Konen als Co-Autor
VIII	1932	654	Konen als Co-Autor;*
VII, Teil 3	1934	723	Konen als Co-Autor;**
* diese erste Lieferung ist zugleich die einzige Lieferung von Band VIII.			
** bei der Erscheinung dieses Teils wurden alle drei Teile von Band VII zusammengefügt.			

In Tabelle 3.2 sind Informationen zu den Rezensionen in bedeutenden Zeitschriften aufgelistet. Es ist zu sehen, dass die meisten Rezensionen aus den USA kamen, gefolgt von Großbritannien. Wie Sommerfeld in der *Zeitschrift für*

Astrophysik erwähnte, stand Kaysers Name besonders bei den amerikanischen Forschern in höchsten Ehren [Sommerfeld (1941)]. Es stellt sich daher die Frage, warum es so wenige deutsche Rezensionen gab? Eine andere interessante Frage ist nämlich, warum schrieb kein deutscher Chemiker eine Rezension, obwohl das Handbuch viele chemierelevante Inhalten enthält und die Spektralanalyse der Elemente als ein wichtiges Gebiet der Spektroskopie aus Deutschland stammt? Auf eine Diskussion zu diesen Fragen wird in diesem Beitrag verzichtet. Sonst würde das hier den Rahmen sprengen.

Tabelle 3.2:
Informationen zu den Rezensionen in bedeutenden Fachzeitschriften

Jahr	Band	Zeitschrift*)	Rezensent**)	Beruf
1901	I	Nature	Schuster (GB)	Physiker/Astronom
1902	I	APJ	Crew (US)	Physiker/Astronom
1903	II	Nature	Schuster (GB)	Physiker/Astronom
1904	II	APJ	Hale (US)	Physiker/Astronom
1905	III	APJ	Coblentz (US)	Physiker/Chemiker
1905	III	Nature	Baly (GB)	Chemiker
1908	IV	APJ	Wood (US)	Physiker/Astronom
1908	IV	Nature	Anonym	–
1911	V	APJ	Crew (US)	Physiker/Astronom
1911	V	Nature	Baly (GB)	Chemiker
1913	VI	APJ	Crew (US)	Physiker/Astronom
1924	VII, 1. Teil	APJ	Frost (US)	Physiker/Astronom
1933	VIII	Nature	Johnson (GB)	Physiker
1933	VIII	JPC	Baly (GB)	Chemiker
1933	VIII	ZAMM	Simson (DE)	Physikerin
1934	VIII	APJ	Morgan (US)	Physiker, Astronom
1934	VII, 3. Teil	ZAMM	Anonym	–

* APJ = Astrophysical Journal; JPC = Journal of Physical Chemistry; ZAMM = Zeitschrift für angewandte Mathematik und Mechanik

** Das Länderkennzeichen hinter einem Namen deutet auf die Staatsangehörigkeit dieses Rezensenten hin.

3.4 Schlusswort

Die rasche Entwicklung der Spektroskopie seit der Entdeckung der Spektralanalyse machte die Verfügbarkeit eines zuverlässigen und zugleich umfassenden Nachschlagwerks auf diesem Gebiet erforderlich. Heinrich Kaysers von 1900 bis 1934 erschienenes achtbändiges *Handbuch der Spectroscopie* erfüllte im Wesentlichen diesen Bedarf. Da die Spektroskopie mit der Astrophysik eng verbunden ist, profitierten auch zeitgenössische Astrophysiker in hohem Maß davon. Dies spiegelt sich in den von bedeutenden Wissenschaftlern geschriebenen Rezensionen in namhaften Fachzeitschriften wieder. Somit trug das Handbuch besondere Leistungen zur Entwicklung der Astrophysik im 20. Jahrhundert bei.

3.5 Literatur

ANONYM: Handbuch der Spectroscopie. In: *Nature* **78** (1908), S. 338–339.

ANONYM: Handbuch der Spectroscopie. In: *Zeitschrift für Angewandte Mathematik und Mechanik (ZAMM)* **14** (1934), 4, S. 255.

BALY, EDWARD CHARLES: Handbuch der Spectroscopie. In: *Nature* **72** (1905), S. 627–628.

BALY, EDWARD CHARLES: Handbuch der Spectroscopie. In: *Nature* **86** (1911), S. 40.

BALY, EDWARD CHARLES: Handbuch der Spectroscopie. In: *Journal of Physical Chemistry* **37** (1933), 4, S. 539–540.

COBLENTZ, WILLIAM: Handbuch der Spectroscopie. In: *Astrophysical Journal* **22** (1905), S. 281–283.

CREW, HENRY: Handbuch der Spectroscopie. In: *Astrophysical Journal* **15** (1902), S. 150–154.

CREW, HENRY: Handbuch der Spectroscopie. In: *Astrophysical Journal* **33** (1911), S. 87–89.

CREW, HENRY: Handbuch der Spectroscopie. In: *Astrophysical Journal* **38** (1913), S. 126–128.

DÖRRIES, MATTHIAS & KLAUS HENTSCHEL (Hg.): *Kayser, Heinrich: Erinnerungen aus meinem Leben [1936].* München: Deutsches Museum (Algorismus; Vol. 18) 1996.

JOHNSON, RAYNOR: Handbuch der Spectroscopie. In: *Nature* **131** (1933), S. 824.

DEVORKIN, DAVID: Astrophysics. In: LANKFORD, JOHN (Hg.): *History of Astronomy – An Encyclopedia.* New York, London: Garland Publishing 1997, S. 72–80.

FROST, EDWIN: Handbuch der Spectroscopie. In: *Astrophysical Journal* **59** (1924), S. 192–194.

HALE, GEORGE: Handbuch der Spectroscopie. In: *Astrophysical Journal* **15** (1904), S. 296–300.

KIRCHHOFF, GUSTAV & ROBERT BUNSEN: Chemische Analyse durch Spectralbeobachtungen. In: *Annalen der Physik und Chemie* **110** (1860), 6, S. 161–189.

SCHUSTER, ARTHUR: Handbuch der Spectroscopie. In: *Nature* **63** (1901), S. 317–318.

SCHUSTER, ARTHUR: Handbuch der Spectroscopie. In: *Nature* **67** (1903), S. 265–266.

SIMSON, CLARA VON: Handbuch der Spectroscopie. In: *Zeitschrift für Angewandte Mathematik und Mechanik (ZAMM)* **13** (1933), 3, S. 248.

SOMMERFELD, ARNOLD: Heinrich Kayser. In: *Zeitschrift für Astrophysik* **20** (1941), S. 308–309.

KAYSER, HEINRICH: Bestimmung des Verhältnisses der specifischen Wärmen für Luft bei constantem Druck und constantem Volumen durch Schallgeschwindigkeit. In: *Annalen der Physik und Chemie* **2** (1877), 3, S. 218–241.

KAYSER, HEINRICH: Ueber den Einfluss der Intensität des Schalles auf seine Fortpflanzungsgeschwindigkeit. In: *Annalen der Physik und Chemie* **6** (1879), 4, S. 465–485.

KAYSER, HEINRICH: *Lehrbuch der Spektralanalyse.* Berlin: Springer 1883.

KAYSER, HEINRICH: Notiz zu den Spectren der Cometen. In: *Astronomische Nachrichten* **134** (1894a), S. 353–356.

KAYSER, HEINRICH: Ueber den Einfluss der Spaltweite auf das Aussehen der Cometenspectra. In: *Astronomische Nachrichten* **135** (1894b), S. 1–10.

KAYSER, HEINRICH: *Handbuch der Spectroscopie, Band I.* Leipzig: S. Hirzel 1900.

KAYSER, HEINRICH: *Handbuch der Spectroscopie, Band II.* Leipzig: S. Hirzel 1902.

KAYSER, HEINRICH: Ziele der Zeitschrift. In: *Zeitschrift für wissenschaftliche Photographie, Photophysik und Photochemie* **1** (1903), S. 1–4.

KAYSER, HEINRICH: *Handbuch der Spectroscopie, Band III.* Leipzig: S. Hirzel 1905.

KAYSER, HEINRICH: *Handbuch der Spectroscopie, Band IV.* Leipzig: S. Hirzel 1908.

KAYSER, HEINRICH: *Handbuch der Spectroscopie, Band V.* Leipzig: S. Hirzel 1910.

KAYSER, HEINRICH: *Handbuch der Spectroscopie, Band VI.* Leipzig: S. Hirzel 1912.

KAYSER, HEINRICH: *Handbuch der Spectroscopie, Band VII, Teil 1.* Leipzig: S. Hirzel 1924.

KAYSER, HEINRICH: *Tabelle der Schwingungszahlen der auf das Vakuum reduzierten Wellenlängen zwischen 2000 Å und 1000 Å.* Leipzig: S. Hirzel 1925.

KAYSER, HEINRICH: *Tabelle der Hauptlinien der Linienspektra aller Elemente nach Wellenlänge geordnet.* Berlin: Springer 1926.

KAYSER, HEINRICH: *Handbuch der Spectroscopie, Band VII, Teil 2.* Leipzig: S. Hirzel 1930.

KAYSER, HEINRICH: *Handbuch der Spectroscopie, Band VIII.* Leipzig: S. Hirzel 1932.

KAYSER, HEINRICH: *Handbuch der Spectroscopie, Band VII, Teil 3.* Leipzig: S. Hirzel 1934.

KAYSER, HEINRICH: Statistik der Spektroskopie. In: *Physikalische Zeitschrift* **39** (1938), S. 466–468.

WOLFSCHMIDT, GUDRUN: Kosmochemie – Chemische Elemente im Kosmos – Meteoriten, Sterne, Kosmologie. In: WOLFSCHMIDT, GUDRUN (Hg.): *Kosmochemie – Geschichte der Entdeckung und Erforschung der chemischen Elemente im Kosmos.* Hamburg: tredition (Nuncius Hamburgensis; Band 50) 2023, S. 67–135.

Figure 4.1:
The photograph shows Karl Schwarzschild (left) and Ejnar Hertzsprung (right), dressed in their purple bordered gowns, so-called “Talare”, in front the entrance of the Gauß Observatory in Göttingen

Cod. Ms. K. Schwarzschild 23: 1,13
(Nachlass K. Schwarzschild; HANS SUB Universität Göttingen)

Karl Schwarzschild and Ejnar Hertzsprung in Potsdam (1910–1916)

Adriaan Raap (Königsbronn)

Abstract: Karl Schwarzschild and Ejnar Hertzsprung in Potsdam (1910 to 1916)

Both astrophysicists, which came into the world approximately within 24 hours in October 1873, and learned to know each other in the middle of their thirties, by exchanging their astrophotographic results. This article will be concerned mainly with the scientific contributions of Schwarzschild made during his leadership as the director of the Astrophysical Observatory in Potsdam in the period from 1910 until his premature death during the First World War in 1916. Hertzsprung, who did accompany Schwarzschild, since his appointment as assistant professor for astronomy in Göttingen, began an extensive upgrade of the Large Refractor in Potsdam.

In the autumn of 1910, Schwarzschild, while attending the *International Solar Union* meeting in Los Angeles, visited all greater observatories, upon crossing the United States. Especially, after visiting Henry Norris Russell (1877–1957) at the *Princeton University*, he realized that Hertzsprung's findings of the relationship between the luminosity and the color-temperature of stars needed recognition by an acknowledged publication to obtain the priority.

Already in September 1914, Karl Schwarzschild and Reinhard Süring, due to their profound knowledge of meteorology and the navigation of balloons and Zeppelins, were ordered to establish a weather station in Belgium. Thereafter, from July 1915, Schwarzschild served as calculator for heavy artillery in the Staff of major-general Johannes von Schadel, respectively, in Russia and in France. At the same time, he stayed always in contact with the works of Albert Einstein on *Special Relativity* and of Arnold Sommerfeld on *Quantum Mechanics.* Then, suddenly at the age of 42 only, Schwarzschild suffered from an auto-immune disease of the skin and soon died in May 1916.

Zusammenfassung: Karl Schwarzschild und Ejnar Hertzsprung in Potsdam von 1910 bis 1916)

Beide Astrophysiker erblickten das Lebenslicht fast gleichzeitig im Oktober 1873, und lernten einander dann erst persönlich kennen, nachdem sie sich regelmäßig über ihre astrophotographischen Ergebnisse austauschten. Dieser Beitrag beschäftigt sich aber vornämlich mit den wissenschaftlichen Arbeiten Schwarzschilds, während seiner Zeit als Geheimrat und Direktor des Astrophysikalischen Observatorium in Potsdam ab 1910, bis zu seinem vorzeitigen Tod 1916 im Ersten Weltkrieg. Hertzsprung, der seit seiner Ernennung als Extraordinarius in der Astronomie an der Universität in Göttingen 1909, Schwarzschild nach Potsdam begleitet hat, verbesserte dort die Leistungsfähigkeit des Großen Refraktors entscheidend.

Im Herbst 1910, anlässlich der Versammlung der *International Solar Union* in Los Angeles, durchquerte Schwarzschild den Vereinigten Staaten, und besuchte dabei alle größere Observatorien. Insbesondere, nachdem er Henry Norris Russell (1877–1957) an der *Princeton Universität* begegnet war, dass Hertzsprungs Ergebnisse, welche auf eine Beziehung zwischen der Leuchtkraft und der Farbtemperatur eines Sterns deuteten, nur Priorität bekommen konnte, wenn eine Publikation in eine anerkannte Zeitschrift bald erfolgte.

Bereits im September 1914, bekamen Karl Schwarzschild und Reinhard Süring wegen ihrer Erkenntnisse in der Meteorologie und der Navigation von Ballonen und Zeppelinen, den Auftrag, um in den belgischen Ardennen eine Wetterstation zu errichten. Anschließend, ab Mitte Juli, wurde Schwarzschild dem Stab des Major-Generals Johannes von Schadel zur Erstellung von Schusstabellen der Schwere Artillerie zugeteilt, und begleitete dessen Stab nach Russland und Frankreich. Gleichzeitig blieb er jedoch auch in Verbindung mit den Arbeiten von Albert Einstein über die *Spezielle Relativitätstheorie* und von Arnold Sommerfeld über die *Quantenmechanik.* Plötzlich im Alter von nur 42 Jahren, litt er an einer Autoimmunkrankheit der Haut und verstarb kurz darauf im Mai 1916.

4.1 Introduction

A comparison of the personalities of Karl Schwarzschild and Ejnar Hertzsprung reveals their totally different characters. Karl Schwarzschild, apart from all social contacts with the members of his large family, also participated in many societies with his colleagues at the university. Moreover, he was engaged in several sporting activities such as horse riding, playing tennis, alpinism, and skiing. Furthermore, he also showed scientific interest in the navigation of Balloons and Zeppelins and made several balloon-flights from Göttingen. Ejnar Hertzsprung, instead, abhorred sports and avoided social obligations as much as possible. Livelong, he liked to spend many hours sitting behind his comparator instrument to evaluate the numerous astro-photographical plates. This behavior brought him and his wife very often in conflict with the regular invitations they got from the "Telegraphenberg"- community in Potsdam.

When they arrived 1910 in Potsdam, both men were at the summit of their creativity, which is proven by the number of astronomical concepts being developed by them in those years, and finally were named after them.

Scientific concepts or expressions based on the works of Karl Schwarzschild are:

- Schwarzschild *exponent*: determines the blackening of photographic plates.
- Schwarzschild *effect*: due to long-time exposures of photographic plates.
- Schwarzschild *geometry*: describes spacetime surrounding a spherical mass.
- Schwarzschild *geodesics*: worldline solutions of the spacetime metric.
- Schwarzschild *radius*: determines the event-horizon of black holes.
- Schwarzschild *criterion*: determines convective stability of a stellar medium.

Nowadays, a lot of synonyms of the above-mentioned concepts are in use, such as:

- Schwarzschild *black hole.*
- Schwarzschild *spacetime.*
- Schwarzschild *cosmology.*
- Schwarzschild *universe.*

although these concepts were never part of a publication by Schwarzschild.

On the far side of the moon, a large impact crater about 210 km in diameter, has been named after Karl Schwarzschild. It is located at the far northern part of the hemisphere. Furthermore, the asteroid *837 Schwarzschilda* is named in Karl's honour.

Scientific concepts or expressions based on the works of Ejnar Hertzsprung are:

- *Hertzsprung-Russell-Diagram: Scatterplot of stars showing the relationship of the luminosity (absolute magnitude) and the effective temperature (classification).*
- *Hertzsprung Gap: A Feature in the Hertzsprung-Russell-Diagram.*

In Potsdam, Hertzsprung began to measure the parallax of several *Cepheid variable stars*, which allowed him to calibrate the relationship between the luminosity and the periods of those stars, discovered by Henrietta Leavitt (1868–1921), and could determine the distance of the *Small Magellanic Cloud.*[1]

On the far side of the moon, an even larger impact crater (diameter 570 km) is named after Hertzsprung. It is located at the equator of the moon beyond the western limb. Also, the asteroid *1693 Hertzsprung* was named after Ejnar in his honour.

Figure 4.2:
Both astronomers married in the period they knew each other personally

Bild links: Herrmann: Ejnar Hertzsprung, 1994.
Bild rechts: Cod. Ms. K. Schwarzschild 23: 1,11 (HANS SUB Göttingen)

1 Hertzsprung, Ejnar: Über die räumliche Verteilung der Veränderlichen vom δ-Cephei-Typus, (1913).

In 1909, Karl Schwarzschild married Else Rosenbach (1879–1950), the eldest daughter of his physician. They had three children Agathe (1910), Martin (1912) and Alfred (1914).

In 1913, Ejnar Hertzsprung married Hetty Kapteyn (1881–1956), the youngest daughter of the astronomer Jacobus Kapteyn (1951–1922). They had one daughter named Rigel. The marriage was divorced after 7 years.

4.1.1 The Scientific Curriculum Vitae of Karl Schwarzschild (1873–1916)

Karl Schwarzschild was the eldest son of Moses Martin Schwarzschild and Henrietta Sabel, and a descendant from the large Jewish Schwarzschild-Dynasty in Frankfurt on Main. He had five younger brothers and a sister, of which during his lifetime he cared much for his brother Alfred Schwarzschild (1874–1948), who was a known artist in portrait-paintings.

Listing of the main stages of his scientific career in astrophysics:

- 1885–1891: First scientific papers at the “Lessing-Gymnasium” Frankfurt.
- 1891–1893: Student at the “Kaiser-Wilhelm” University Strasbourg.
- 1893–1897: Graduate student at the “Ludwig-Maximilian” University Munich.
- 1897–1899: Assistant at the “von Kuffner” Observatory Vienna-Ottakring.
- 1899–1901: Habilitation at the “Ludwig-Maximilian” University Munich.
- 1901–1909: Professor at the “Georg-August” University Göttingen.
- *1909–1916: Director of the Astrophysical Observatory Potsdam.*
- 1914–1916: Officer in the German (Prussian) Army (Belgium, France, Russia).

In Göttingen, Schwarzschild, like its first director Carl Friedrich Gauß (1777–1855), worked and lived in the Observatory.

4.1.2 The Scientific Curriculum Vitae of Ejnar Hertzsprung (1873–1967)

Ejnar Hertzsprung was the first-born child of Severin Carl Ludvig Hertzsprung (1839–1893) and his wife Henriette Christiane Charlotte Frost. Moreover, Ejnar had a brother named Ivar Hertzsprung (1875–1902), who died at young age, and a sister Ellen (1879–1963), who stayed unmarried.

Listing of the main stages of his scientific career in (photo-)chemistry and astronomy:

- 1892–1898: Student in chemistry at the Metropolitan-School Copenhagen.
- 1898–1899: Large interest in Stereophotography (Amateur photographer).
- 1899–1901: Engineer at the Carbid and Acetylen Company St. Petersburg.
- 1901–1902: Study of Photochemistry at Wilhelm Ostwalds Laboratory in Leipzig.
- 1902–1909: Astronomy at the Urania-Observatory and University Copenhagen.
- 1909: Associate Professor at the Georg-August University Göttingen.
- *1909–1919: Observer of the Astrophysical Observatory Potsdam.*
- 1919–1944: (Co-)director of the Observatory – University of Leiden.
- 1945–1967: Observatory of Brorfelde – University of Copenhagen.

The appointment of Ejnar Hertzsprung in 1909 as an associate professor ("Extraordinarius") in Göttingen *"in the twinkling of an eye"*, as Karl Schwarzschild did describe this appointment, was possible due to the fact, that the mathematician Gustav Herglotz (1881–1953) had just accepted a new position at the University of Vienna. Moreover, Schwarzschild had a good relationship to the dean of the Georg-August University, Carl Runge (1856–1927), since their common solar-eclipse expedition to Guelma (Algeria), 1905, August 29–30. Furthermore, in order that Schwarzschild also could take Hertzsprung with him to Potsdam, he succeeded in getting an exchange with Johannes Hartmann (1865–1936).

4.2 Director of the Astrophysical Observatory in Potsdam

Originally founded in 1874, this institution was equipped with several types of telescopes used for the cataloguing of stars (projects: *"Carte du Ciel"* and *"Bonner Durchmusterung"*). These telescopes, however, were of limited use in determining the radial velocities of stars, which is based on photographic spectroscopy. The leading position of the Astrophysical Observatory in Germany and abroad was endangered. Therefore, at the beginning of 1895 design studies were made by the known instrument making and optics company Repsold and Steinheil for the building of a new telescope, which should allow for excellent spectroscopic research of stars in future. This was one reason that Hertzsprung and Schwarzschild were very eager in going to work with the Great Refractor, the largest telescope in Potsdam, in August 1909.

Karl Schwarzschild, however, who just got married to Else Rosenbach (see above) from Göttingen, left this university town with all his friends and colleagues, heavy-heartedly.

4.2.1 Problems with the appointment of Karl Schwarzschild

From the beginning the appointment of Karl Schwarzschild as the new director of the Astrophysical Observatory in Potsdam wasn't undisputed. Already, throughout the discussions about the successor of its first director Hermann Carl Vogel (1841–1907) differences did turn out between the professional astronomers [Arthur Auwers (1838–1915); Gustav Müller (1851–1925)] and the theoretical physicists of the *Prussian Academy of Science* [for instance Max Planck (1859–1947); Albert Einstein (1879–1955)].

Auwers pleaded for the many years of observation made at the Astrophysical Observatory in Potsdam, remarking, that Müller would guarantee *"a quite homely, but nonetheless solid, conscientious and elaborate work at the AOP."*

Karl Schwarzschild instead, although being a graduate student of Hugo von Seeliger (1849–1924), wasn't regarded as a real astronomer at all, but more or less as a theoretical physicist. At that time the physicists in Berlin, especially Max Planck, got through their recommendation, and Karl Schwarzschild became the new AOP-director against Auwer's vote.

In the background, moreover, a decisive role in Schwarzschild's appointment was played by the mathematician Felix Klein (1849–1925), in fact as a continuation of the "System Althoff", which for instance made the Georg-August University of Göttingen to an international leading center of mathematics and physics),[2] until it became destroyed in the National Socialists period.

2 Als System Althoff wir das unbürokratische und oft die Ressortgrenzen überschreitende Vorgehen des preußischen Friedrich Althoff (1839–1908) bezeichnet (siehe Wikipedia).

4.2.2 The Great Double Refractor – Combined guiding and photographic telescopes

The Great Refractor in Potsdam was constructed as a parallactic double telescope mounted equatorially inside a large cylindrical building with a rotatable dome (see Fig. 4.3). The smaller guiding telescope was equipped with a 50 cm objective lens and a focal length of 12.59 m, which was mainly intended for the visual observations. The larger photographic telescope possessed an 80 cm objective and a focal length of 12.14 m, which at the end could be connected to spectrographic instruments, specially developed for the double refractor. The spectrograph made by Steinheil (31 kg) in Munich possessed three prisms, a second spectrograph made by Carl Zeiss (20 kg) in Jena had only one prism.

The new building and instruments on the Telegraphenberg were inaugurated in August 1899 in the presence of the German Emperor and King of Prussia Wilhelm II (1859–1941).

Here, from the beginning, Johannes Hartmann soon discovered that the objective lens of the larger telescope showed astigmatism and chromatic aberration. To this aim he developed an opaque screen containing a series of tiny holes covering the aperture of the telescope and used it for testing the focal points of the objective lens at different wavelengths, taking a series of "spot diagrams". Thus, he determined, that the problems resided in the primary lens, which was readjusted under his supervision.

Further adjustments of the Great Refractor lenses were undertaken by Schwarzschild and Hertzsprung when they came to Potsdam in August 1909. They entrusted the Estonian optician Bernhard Schmidt (1879–1935) with the correction and improvement of the 50 cm objective lens of the Steinheil visual refractor in 1911 and again in 1914.

Although Schwarzschild worked untiringly on his own research and that of his collaborators, the leadership of this remarkable institute took him much time expenditure with all kinds of non-scientific activities.

4.2.3 The Comet-Fever of the year 1910

The *Great January Comet of 1910*, also called the *Daylight Comet* (ref.: C/1910 A1), which at first was observed at the Southern Hemisphere, originally was held for an early return of *Halley's Comet.* The former Comet, therefore, came as something as a surprise, and was visible in South Africa to the naked eye at dawn. There it was observed scientifically by the astronomer Robert T. A. Innes (1861–1933) from the *Transvaal Observatory* in Johannesburg on January 17. After passing its perihelion, the Great January Comet moved northwards

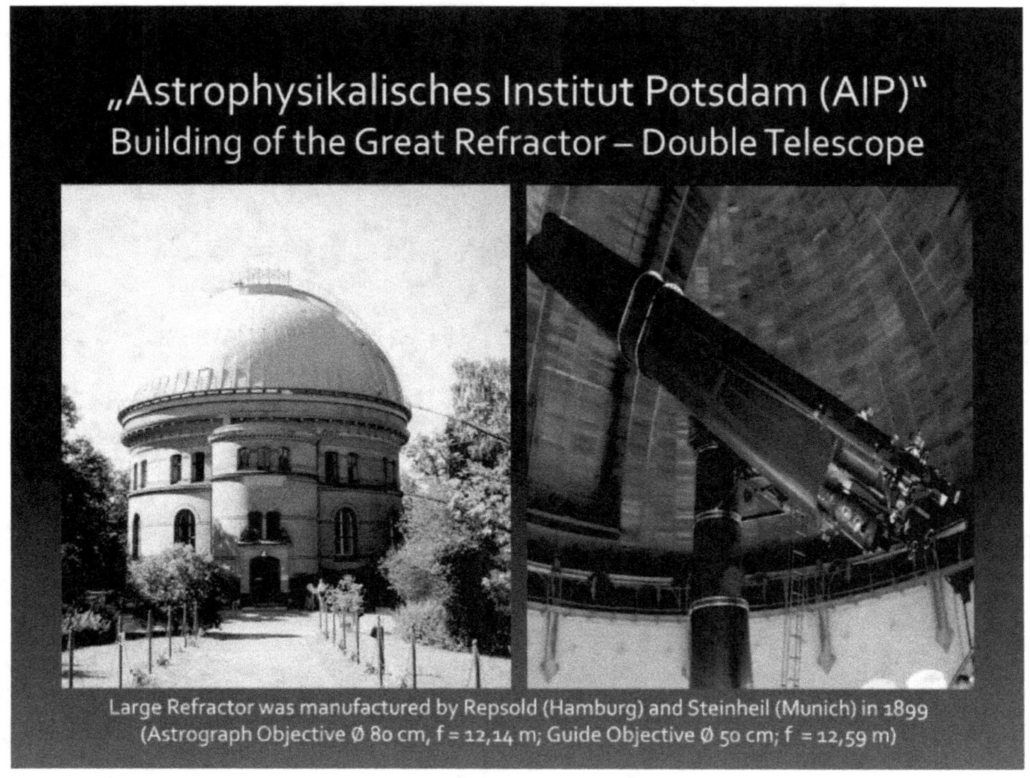

Figure 4.3:
Recent photographs of the Great Refractor Building and of the Great Refractor, now located in the Albert Einstein Science Park (i. e. the former "Telegrafenberg") in Potsdam

and became visible at the Northern Hemisphere even by day. The enormous brightness of this Comet (max. mag. −5) outreached that of the Halley Comet (max. mag. −2), which reached its perihelion on April 20. At it brightest, both comets outshone the planet Venus, and were possible the brightest comets of the 20th century.

The appearance of both comets caused a widespread panic across the world. The French astronomer and popular scientist Camille Flammarion (1842–1925) did warn that the Earth would pass through the comet's tail in May 1910. Flammarion was widely reported in numerous American newspapers, saying that there was a chance that the highly toxic Cyanogen gas would impregnate

Figure 4.4:
Left: The *Daylight or Sunset Comet*, photo: Percival Lowell
Right: *Halley-Comet* of May 29, 1910, photo: Edward Emerson Barnard

Bild links: Great January Comet of 1910 (C/1910 A1), (photo: Lowell Observatory, CC)
Bild rechts: Halley's Comet (P1/Halley), (photo: Yerkes Observatory, CC)

the Earth atmosphere and possibly *"extinguish all life on the planet."*[3] In fact, he said no such thing. Once, announced by the mainstream media, this caused panic and many feared that the end of the world was approaching. This

3 During the appearance of the Morehouse Comet (C/1908 R1) in September 1908, the astronomers at the *Yerkes Observatory* in Williams Bay analysed spectroscopically the composition of its tail and found high concentrations of poisonous CO^+ and Cyanogen ($N \equiv C - C \equiv N$) gas. Two years later the scientists calculated that the Earth would pass directly through comet's Halley's tail and its gas would saturate the atmosphere, killing all life on Earth.

fueled swindlers to sell *"Comet fever pills of Dr. Halley"* to ward off *"cometary poison"*, gas masks, etc.

4.2.4 The empress and the crown princess visit the Great Refractor

The excitement caused by the appearance of two comets early 1910, which even in a large city as Berlin must have been clearly visible in the sky, although well before 1900 the astronomers already had started to complain about the light-pollution at night in the larger university cities. Anyhow, the comets were also visible by day, so that the empress Augusta Viktoria (1858–1921) and her daughter Viktoria Luise (1892–1980), while staying at their "Potsdamer" residence, spontaneously took the opportunity to visit the Great Refractor at the "Telegraphenberg". Karl Schwarzschild and his collaborators must have been very surprised by the visit of this very highest society. In a letter of to his parents he reported:

> Citation (translation): "[dots] *The very highest visit came all of a sudden. Surely, I made many horrible mistakes and said You ("SIE") to the royal highnesses. The crown princess is clever and charming and understood what was show, the others looked around and did as if they were interested."* [dots]
> *"The empress went on her knees without problems to look in the telescope, what amazed me of this elder woman. She made the impression, as if she did not understand a word. Finally, everybody of the imperial family gave me a hand and thanked me deeply – so that at last, I noticed who belonged to them."* [dots]
> *"It all passed over like a ghost story."*[4]

4.3 Karl Schwarzschild as Balloon- and Zeppelin-aeronaut

In Göttingen, Karl Schwarzschild became a member of the local Balloon society, in which also several colleagues from the Georg-August-University, such as Professor Ludwig Prandtl (1875–1953), the father of aerodynamics, and Professor Carl Runge (1856–1927), a specialist in solving the non-linear differential equations of dynamical systems, participated.

4 Schwarzschild-Nachlass: SUB/Briefe 907:27/1910-05-XX/Potsdam.

Figure 4.5:
The empress Auguste Victoria von Schleswig-Holstein with her daughter Viktoria Luise of Prussia during a trip in Berlin, 31 May 1911

Bundesarchiv, Bild 102-00621 / CC-BY-SA 3.0, Georg Pahl

4.3.1 Astronomical navigation with a Balloon-Sextant at night

The astronomical navigation at the beginning of the twentieth century by day or by night are different. At night this can be achieved most effectively by measuring the height of two bright stars to determine the latitude and longitude of the location. Furthermore, the values of both star altitudes should be readily transferred into the local coordinates with the help of star-tables. These tables were provided by detailed calculations of Karl Schwarzschild and Dr. Otto Birck (1879–1951).[5]

Compared with a common sextant, in which the images of the star and the reflection of the spirit-level are combined by an adjustable mirror into field of

5 Schwarzschild & Birck: Tafeln zur astronomischen Ortsbestimmung im Luftballon, 1909.

the telescope, the Schwarzschild-sextant instead is equipped with a spirit-level, which is fixed parallel to the telescope; the image of the star is then brought by a rotating mirror and another reflection by a fixed mirror into a position, which covers the middle of the bubble image of the spirit-level into the telescope.[6]

Another very efficient and easy method for transforming the observed heights of stars into the location of a balloon at night, which was especially useful for non-astronomers, were the tables by Schwarzschild and Birck, only based on the observations of the polar star, a star in the east and one in the west. No use has been made of the sideral time and other astronomical terminology.[7]

Other contributions related to the navigation of balloons were published by Karl Schwarzschild.[8]

4.3.2 The *"Internationale Luftschiffahrt-Ausstellung"*

On the occasion of the very first *"Internationale Luftschiffahrt-Ausstellung (ILA)"*[9] which was held in Frankfurt am Main from July until October 1909, Karl Schwarzschild made a flight from Frankfurt am Main to Düsseldorf, testing his navigation instrument while sitting on top of the Zeppelin.

4.4 The large Observatories of the United States

In the autumn of 1910, Karl Schwarzschild travelled to the United States of America, to participate the meeting of the *International Solar Union* in Pasadena. On his way from the East- to the West-coast he visited all larger observatories of the USA. At the East-coast these were the observatories of the *US Naval Observatory* in Washington D. C., the *Harvard College* in Cambridge (Massachusetts) and the *Yerkes Observatory*, in Williams Bay (Wisconsin). After visiting the Niagara Waterfalls at the USA/Canadian border, he travelled by train to the West-coast, where at first he visited the Grand Canyon National Park in Arizona, before going to the *Mount Wilson Observatory* and the *Lick Observatory* at Mount Hamilton in California.

6 Lepsius & Wachsmuth: Denkschrift der ersten internationalen Luftschiffahrts-Austellung (ILA) zu Frankfurt am Main, 1909. Berlin 1910, p. 107–108.

7 Lepsius & Wachsmuth 1910, p. 109.

8 Schwarzschild, Karl: [dots] Ortsbestimmung im Luftballon, (1910), p. 75–80, 204 (note). Schwarzschild: Künstlicher Horizont und Ballonsextant, (1910), p. 357–359. Schwarzschild: Libellenhorizont und Libellensextant, (1913), p. 177–180.

9 Lepsius & Wachsmuth: Denkschrift der ersten internationalen Luftschiffahrts-Austellung, Berlin 1910, p. 106.

Figure 4.6:
Balloon Sextant according to a design of Karl Schwarzschild, manufactured by the Company Spindler & Hoyer of Göttingen

(Collection of historical instruments in the Institute of Astrophysics, Georg-August-University, Göttingen)

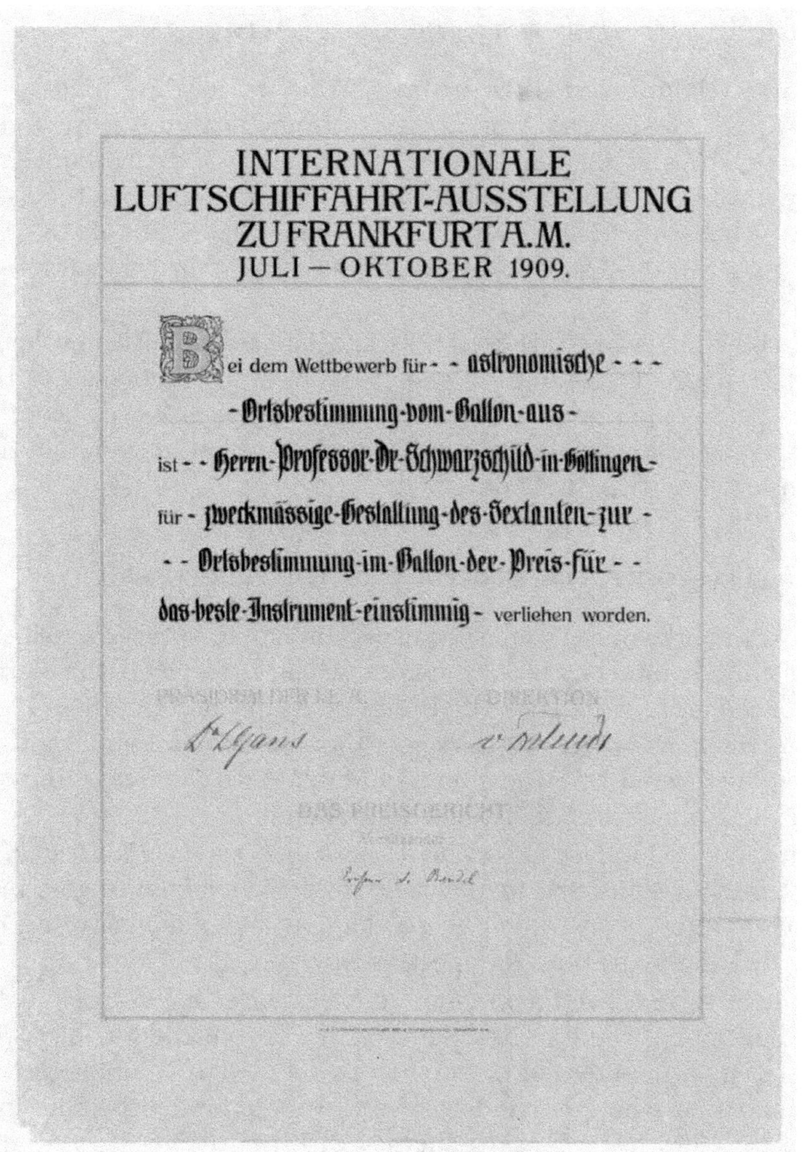

INTERNATIONALE
LUFTSCHIFFAHRT-AUSSTELLUNG
ZU FRANKFURT A.M.
JULI – OKTOBER 1909.

Bei dem Wettbewerb für - - astronomische - - -
- Ortsbestimmung vom Ballon aus -
ist - - Herrn Professor Dr. Schwarzschild in Göttingen -
für - zweckmässige Gestaltung des Sextanten zur -
- - Ortsbestimmung im Ballon der Preis für - -
das beste Instrument einstimmig - verliehen worden.

Figure 4.7:
The content of this document reads as follows:
During the competition of *"astronomical localization from a balloon Professor Dr. Schwarzschild in Göttingen"* has been given *"for his effective sextant-modification to determine the position of a balloon the prize of the best instrument"* Praesidium of ILA: Dr. L. Gans; Direction: v. Tschudi; Prize Commision: Prof. Dr. Brendel.

(Cod. Ms. K. Schwarzschild 22 (Nachlass K. Schwarzschild; HANS SUB)

4.4.1 Atlantic Crossing with the s. s. "AMERICA"

Karl Schwarzschild, boarded the steamer "America" of the *Hamburg-America-Line* (HAL),[10] crossing the Atlantic Ocean from Hamburg to New York, with intermediate stops at the harbours of Southampton and Cherbourg, from the 28th July until 6th August 1910. On board, also were present the astronomer Hermann von Struve (1854–1920) and his wife Olga (1864–1918). They all stayed together in the Astor Hotel, at Times Square, when they arrived in New York.

Amongst his luggage, Karl Schwarzschild had taken with him his balloon-sextant (see above Fig. 4.6), *"as a toy"*. Karl's measurements of the ship's position with his apparatus appeared to be more accurate, as those made by the crew. Namely, due to the mist at the horizon the ship could not determine its length for more that three days.[11]

4.4.2 Observatories and Institutes at the Eastcoast

Leaving New York by the Pennsylvania Railroad, Karl Schwarzschild arrived alone in the US-capital city of Washington, where he stayed in the Metropolitan Hotel. For dinner he was invited by Professor Asaph Hall jr. (1859–1930) of the *US-Naval Observatory.* He was the son of the astronomer Asaph Hall sr. (1829–1907), who discovered 1877 the moons of Mars. With the Great Refractor they observed the Uranus-moons: Titania and Oberon.

In Baltimore, Schwarzschild made a short visit to the *Physical Institute* of the *John Hopkins University* in Maryland, directed by John August [Anderson] (1873–1959), who showed him the production and manufacturing of highly resolving diffraction-gratings for spectroscopy.

Back again to Hotel Astor in New York city after having visited Washington and Baltimore, where he had suffered much from the terrible hot climate of those cities, Schwarzschild decided to travel next to Cambridge by taking the steamboat on the cool Hudson-river. In Cambridge he was invited by both Pickering-brothers: Edward Charles Pickering (1846–1919) and William Henry Pickering (1858–1938) astronomers at the *Harvard-College Observatory.* Schwarzschild reports that he and Oskar Backlund (1846–1916), a Swedish-Russian astronomer were both invited to participate at the meetings and social activities of the *American Astronomical Society.*[12]

10 *Hamburg-Amerikanische Packetfahrt-Actien-Gesellschaft* (HAPAG), founded in 1847.

11 Schwarzschild-Nachlass: SUB/Briefe 909:3/1910-08-01/23-26/Dampfschiff "AMERIKA"-HAL.

12 Schwarzschild-Nachlass: SUB/Briefe 909:3/1910-08-21/82/Niagara Falls.

Figure 4.8:
The HAPAG-Steamship *Amerika* was a steel-hulled, twin-screw, passenger liner, built for Transatlantic-crossings by the Hamburg-America-Line (HAL).

(Amerika – Schiff, Hamburg-Amerika-Line, 1905, CC)

Most importantly, he also had the opportunity in Cambridge to speak over five hours with Miss [Annie Jump] Cannon (1863–1941) and Miss [Henrietta Swan] Leavitt (1868–1921), showing him their excellent work on variable stars (Cepheids) and on their classification of stellar spectra.[13]) Unfortunately, what exactly they did discuss, has not been reported by Karl Schwarzschild. But it may be assumed, that at least, he brought into their attention the observations and star-classifications, which Ejnar Hertzsprung had published in

13 Schwarzschild-Nachlass: SUB/Briefe 909:3/1910-08-18/97/Cambridge.

two relatively unknown articles: *"Zur Strahlung der Sterne"* (On the radiation of stars)[14] in 1905 and *"Zur Strahlung der Sterne II"*[15] two years later. In those publications Hertzsprung concluded that the G-, K- and M-stars may be divided into two groups of different luminosity. Furthermore, he found that bright red stars must be very large, and that the small number of red giants proved their rapid development.

After a short stay at the Niagara-Falls, and a visit to Albert [Abraham] Michelson (1852–1931) of the Physical Institute at the *University of Chicago*, where Michelson showed him his latest interferometric experiments to determine the rotation of the Earth against the Ether, he went to the East Coast by train overnight, accompanied by Pickering, Backlund and Herbert Hall Turner.

4.4.3 Observatories and institutes at the Westcoast

Before the participants of the Solar Union Meeting arrived in Pasadena and were going to the *Mount Wilson Observatory*, several astronomers gathered at the Grand Canyon first. Karl Schwarzschild, accompanied by Philip Fox (1878–1944), went down to the Colorado-river and returned late in the afternoon, both covered by the "Cambrian-Sandstone"! Later, accompanied by Herbert Hall Turner (1861–1930) of the University of Oxford, England, who is credited coining the word *parsec*, Schwarzschild mastered the 1500 m-climb to Mount Wilson in nine hours. At the summit he was welcomed by Jacobus Kapteyn (1851–1922) of the University of Groningen, who possessed a cottage at Mount Wilson with his wife and was invited for having pancakes with them.[16]

On his way to the *Lick Observatory* on Mount Hamilton in California, Schwarzschild stayed several days with his younger brother Otto Schwarzschild (1878–1944), who had emigrated to the United States and lived at that time in San Francisco. This city center had been destroyed by a heavy earthquake four years before.

4.4.4 The Mount Wilson Conference of the Solar Union, August 1910

This meeting, which was organized by Professor George Ellery Hale (1868–1938) of the *Mount Wilson Observatory*, was attended by 83 international sci-

14 Hertzsprung, Ejnar: Zur Strahlung der Sterne, (1905), p. 429–442.

15 Hertzsprung, Ejnar: Eine Annäherungsformel für die Abhängigkeit zwischen Beleuchtungshelligkeit und Unterschiedempfindsamkeit des Auges, (1907), p. 468–472.

16 Schwarzschild-Nachlass: SUB/Briefe 909:3/1910-08-30/108-109/Mount Wilson.

entists. Many participants, mainly from the United States, had also attended the meeting of the *Astronomical and Astrophysical Society of America*, held before at the *Harvard Collage Observatory*, Cambridge, and thereafter had crossed the continent on a special train, arriving in Pasadena on Sunday, August 28. Of large interest on the next day was a visit to the large machine built to grind the great 100-inch mirror in diameter, 13 inches thick and weighing $4\frac{1}{2}$ tons.

Schwarzschild is proud, that the construction of the new 100-inch mirror at Mount Wilson, is based on the formulas in his articles about geometric optics![17] Due to the health problems of George Ellery Hale at the time of the meeting in Pasadena, Schwarzschild was not able to discuss with Hale details of the construction of the huge 60-inch and 100-inch reflecting telescopes, based both on Schwarzschild's reflecting telescope type.

An outstanding evening lecture, much admired by Karl Schwarzschild, was held by Jacobus Kapteyn: *"Star-Streaming of Stars of the Orion-type"*. Kapteyn's investigations at Mt. Wilson of the motions of the Orion stars proved that they could be divided in two portions moving in different direction, each with about the same velocity. The first group of stars is situated mainly in the Scorpio and Centaurus region and the second group in the Perseus region. The contact with Kapteyn and his family would become very important for both Karl Schwarzschild and Ejnar Hertzsprung. In the absence of an own large observatory in the Netherlands, Kapteyn was allowed to use the *Mount Wilson Observatory*, as well as by his future son-in-law Hertzsprung.

Karl Schwarzschild was chosen as a member of a new *Committee on Stellar Spectra*, which became chaired by Edward Charles Pickering. Professor Pickering had explained, how the notation of the *Draper Catalogue* came about. In the beginning at Harvard, it had been just an arbitrary classification by empirical absorption lines, indicated by letters. As the understanding of the stellar spectra increased many types were dropped except the six denoted by B, A, F, G, K, M.

The entire equipment on the mountain was open to inspection by the delegates. The operation by the *Snow Horizontal Telescope*, including the spectroheliograph, and the Tower Telescope found much interest. The enormous perfection of the 60-inch reflector was demonstrated by the observation many new features in the Hercules Cluster, the Trapezium in the Orion and on the planet Saturn.

Finally, by the initiatives of the physicist Heinrich Kayser (1853–1940), spectroscopist at *University of Bonn* and the astrophysicist Karl Schwarzschild, it

17 Schwarzschild-Nachlass: SUB/Briefe 909:3/1910-09-01/112/Mount Wilson.

was decided that the next meeting of the *Solar Union* should take place in Germany, Bonn, 1913.[18])

4.5 Karl Schwarzschild during the World War I

At the outbreak of the First World War in August 1914, Schwarzschild joined the German Army out of a sense of patriotic duty, as did at that time many scientists, who later were Noble Prize Winners, such as Fritz Haber (1868–1934), James Franck (1882–1864), Gustav Hertz (1987–1975) and Otto Hahn (1879–1968). Exceptions were Albert Einstein and even Ejnar Hertzsprung, who therefore became treated with numerous hostilities by his colleagues Andreas Galle (1858–1943)[19] and Hans Ludendorff (1873–1941)[20] in Potsdam.

From his personnel journal Karl Schwarzschild reported the following about his experiences in the first days in September 1914:

> *"Amused we travelled from Berlin 8 h until Cologne 6h30. Süring*[21] *and I, Corporal Friese, voluntary home guard* ["Landsturm"], *40 years old like me, an artist of the "Königl. Preuß. Meßbild-Anstallt", Corporal Dümke, 25 years old, carpenters-sun,* [dots]."
> *"In Cologne we were welcomed by three very tall officers, who had problems with the transition from a sergeant* ["Feldwebel"] *to a councilor* ["Geheimrat"]. *Our men got bad accommodation in the Zeppelin airship-hangar. Süring and I were in a hotel together with the officers. They all were waiting for leading positions of future Zeppelins and considered it as a pleasant opportunity to drive us around by car."*

18 Chant, Clarence Augustus: The Mount Wilson Conference of the Solar Union, (1910), p. 356–372.

19 Andreas Galle (1585–1943) was a geodesist and a son of the astronomer Johann Galle (1812–1910).

20 HANS Hans Ludendorff (1873–1941) was a younger brother of the General Erich Ludendorff (1865–1937).

21 Reinhard Süring (1866–1950) was an important meteorologist and the director of the *Prussian Meteorological Institute* in Berlin and of the *Magnetic Meteorological Observatory* in Potsdam (1909). On July 31, 1901, Reinhard Süring and Arthur Berson (1859–1942) reached an altitude record by humans of 10,800 m in an open gondola balloon and were the first to detect the stratosphere (by the rising of its temperature).

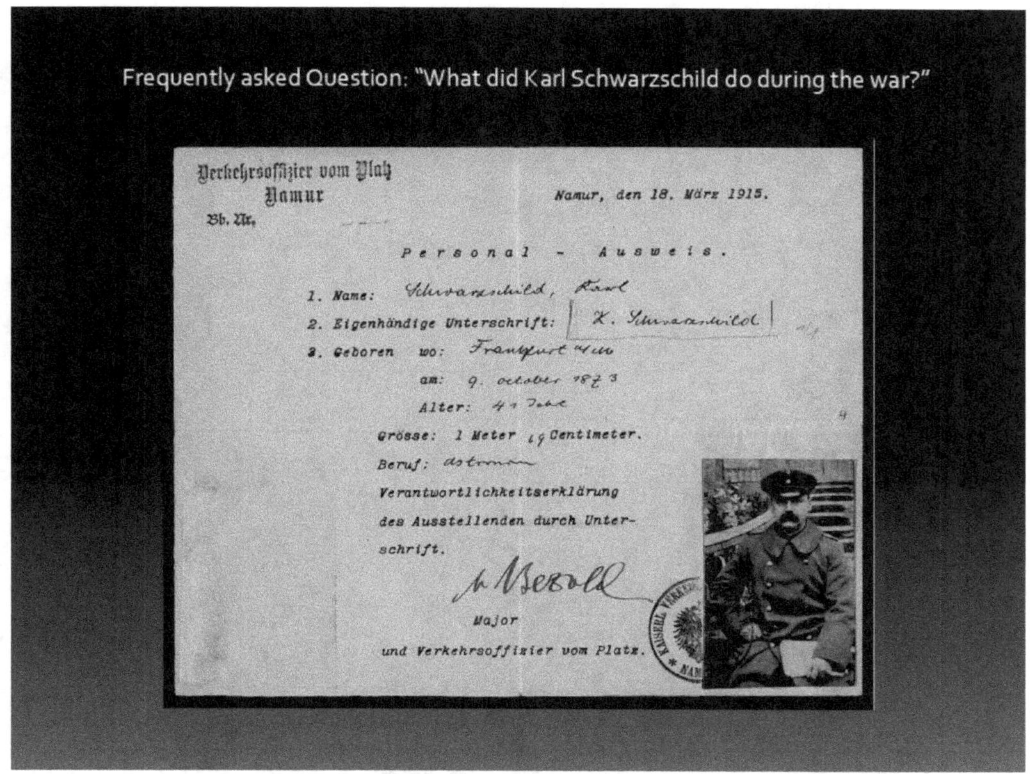

Verkehrsoffizier vom Platz
Namur
Bb. Nr.

Namur, den 18. März 1915.

P e r s o n a l - A u s w e i s .

1. Name: Schwarzschild, Karl
2. Eigenhändige Unterschrift: K. Schwarzschild
3. Geboren wo: Frankfurt a/M
am: 9. october 1873
Alter: 41 Jahr
Grösse: 1 Meter [illegible] Centimeter.
Beruf: Astronom
Verantwortlichkeitserklärung des Ausstellenden durch Unterschrift.

Major
und Verkehrsoffizier vom Platz.

Figure 4.9:
Schwarzschild's Identity Card during World War I *Personal-Ausweis, 18. März 1918*

(Cod. Ms. K. Schwarzschild 22,
Nachlass K. Schwarzschild; HANS SUB Universität Göttingen)

4.5.1 The establishment of a Zeppelin weather station in Namur (Ardennes-Belgium)

After their arrival in Namur, a location for the Zeppelin weather station had to be found. At first, against the will of the Jesuits, Schwarzschild and Süring took the Observatory of the Jesuit College (*Collège Nôtre Dame de la Paix*) into consideration. In fact, the Jesuit College then served as a hospital for the wounded Belgian soldiers. Therefore, they decided to search for an alternative on the citadel, which overlooks the city of Namur:

> "[dots], *immediately again we went up to the Citadel, which towers above the city center of Namur directly by 50 m, looking for a better*

station. It is the castle of the Count of Namur, recently fortified, but already outdated, in there only ammunition and fuel were stored by the Belgians. Tourism plaques in 4 languages informed about its history. At the furthest point a small eight-sided tower exists with glass-windows all around, inside it there were several matrasses, a table, ink, and paper, all just made for us."
"A wonderful place overlooking the City, the Meuse and the Sambre. Directly at the base of the Citadel, Süring and I obtained excellent quarters in a quiet suburb."

Figure 4.10:
An old postcard showing the weather station in Namur during the German occupation of the citadel in World War I (Tour de guetteurs = watch-tower)

(Namur, 1914–1918, Postkarte, private collection)

4.5.2 The Heavy Artillery Staff of Major-General Johannes von Schabel

After finishing the weather station in Namur, Karl Schwarzschild served as a meteorologist and calculator of fire tables for heavy artillery in the Staff

Figure 4.11:
Near Namur, at the fortress of Cognelée, three Zeppelin-hangars were built, designed to bombard Paris and other large cities at the western front.

(Zeppelin-Hangars in Belgien, 1914–1918, Postkarte, private collection)

of Major-General (*"Generalmajor"*) Johannes Alois von Schabel (1857–1945). The latter was a son of the physician in the city-center of Ellwangen, which is situated at the foothills of the Swabian-Alb. Beginning his career in the army of the Kingdom of Württemberg, he was promoted to the Prussian Artillery-Inspection in Berlin. There Johannes von Schabel became involved in the development of the 42-cm Mortar (1910–1913), better known as the *"Dicke Bertha"*, together with Professor Friedrich Rausenberger (1868–1926) at the Krupp AG in Essen.

In January 1915, the Staff of General von Schabel, moved to the Eastern-front for the testing of the so-called *T-Stoff*[22] named after its developer, the chemist

22 In World War I, Major-General Gerhard Tappen (1866–1953) of the Supreme Command (OHL) was responsible for the production of "Shells and Bombs", including chemical warfare. Therefore, he did ask his brother Hans Tappen, a chemist, who proposed to fill artillery shells with the lachrymator xylyl bromide (a tear gas). It did not fulfill the expectations, because the evaporation-rate, especially in wintertime was far too slowly.

Dr. Hans Tappen (1879), near the Russian fortifications around Osowiec. Because of the negative results obtained with this tear gas, it was decided by the Supreme Army Command, that a more lethal gas was needed. Therefore, in April 1915, the Staff of General von Schabel was ordered to return to the Western front near the town of Ypern in Belgium, where Professor Fritz Haber of the *Kaiser Wilhelm Institute for Physical Chemistry and Electrochemistry* had prepared the first chlorine-gas attack at the western-front. After the ending of this Second Battle of Ypres, General von Schabel got the order of the Supreme Army Command to write a manual about the application of special artillery: *"Instructions using gas ammunition"*.

By the middle of August 1915, Karl Schwarzschild joined the Staff of General von Schabel as meteorological assistant and calculator of firing-tables of heavy artillery, in particular for the so-called *"Langer Max"*, which became into their first action during the Battle of Verdun on 21. February 1916 (see below).

Figure 4.12:
The *"Langer Max"* (38-cm-SK-L/45) was a Krupp-development originally built for the Imperial Navy (*"Kaiserliche Marine"*)

(Bundesarchiv, Bild 102-00153 / CC-BY-SA 3.0)

4.5.3 The Head Quarters during the Winter of 1915/16: Mulhouse and Stenay

In December 1915, the Staff of General von Schabel had moved to a new Head Quarter in the center of Mulhouse/Mülhausen. The acts of war here were reduced to artillery duels around the hills of the Hartmannsweilerkopf in the region Vosges. Karl Schwarzschild and the officers lived comfortable in *Villa Jaquet*, which was confiscated from a family of textile-industrialists Eugène (1859–1949) and his son André Jaquet (1885–1959), both had fled to Switzerland before. The parc surrounding the villa with its horse stables invited the officers to ride into the city. In addition, the villa had a billiard and a smoking room, and a library with French literature only, Karl reported to his wife.

Most officers joined their families during Christmas and the New Year; therefore, Karl was nearly left alone in this villa. This was an excellent opportunity for him to concentrate himself fully on his new ideas he had in mind about Einstein's general relativity and Bohr's *quantum theory.* The only person, who did accompany him there, was Alfred Wegener (1880–1930), a meteorologist and polar researcher, who recently had returned from his adventurous Second Greenland Expedition. Wegener told Schwarzschild, how they, with no food left at the end of the crossing of Greenland, by pure luck had survived this expedition. They also discussed about Wegener's *continental-drift theory.*[23]

In January 1916, already, von Schabel's Staff had to move again to Stenay, at the banks of the river Maas, and near the border the Ardennes Forest. Here the Crownprince Wilhelm (1882–1951), as the commander in chief of the 5th Army, had his Head Quarter. However, in March Schwarzschild left this town in Lorraine to recover from his autoimmune illness to Potsdam.

4.5.4 "On the Gravitation on the Battlefield and in Spacetime"

In Mulhouse, Karl Schwarzschild complaints to his wife, that he is becoming more and more impatient, because the war will not end. Instead of mapping the

23 Schwarzschild-Nachlass: SUB/Briefe 909:10/1916-01-03/01-04/Mülhausen. Quotation: *"My loneliness just began in getting too much to me, there appeared Alfred Wegener, who has a brother here, leading a weather station and were 12 hours together. That on the other hand was too much again. But he did learn playing the billiard in Belgium just as I, and we enjoyed it. What he told about his Greenland Expedition, where the glacier beside their camp calved and an icecap with a thickness of 300 m sank into the sea and the icebergs turned over, sank and emerged again, and then, how they had no more food since 31 hours and were sitting at the shore a day's march away from their goal over a mountain ridge, got unconscious every moment and were saved from starving by a ship, which passed purely by chance, that was frightening and yet that much better as all the destructions by war."*

Figure 4.13:
The last known picture showing Karl Schwarzschild at the left
(according to his brother-in-law Robert Emden)

(Cod. Ms. K. Schwarzschild 22, Nachlass K. Schwarzschild; HANS SUB)

results of enemy (and friendly) artillery by the mess squads cartographically, he wants to study the latest articles by Einstein, and Pieter Zeeman (1865–1943).[24] Now he is calculating a trajectory of 38-cm ammunition reaching a distance of 38 km, rising into a height of 11 km, and studies the influence of air pressure and wind, according to his formulas based on the variation theory.[25] The publication of this paper, which was classified as *"very secret"* by the *Oberste Heeresleitung* (OHL = Supreme Army Command), and became first printed in 1920, 2 years after the war.

Einstein's publications in 1915 of his *General Theory of Relativity* allowed him to solve his non-linear field equations by applying an approximation method to determine the perihelium-shift of the planet Mercury. This had raised the

24 Schwarzschild-Nachlass: SUB/Briefe 909:09/1915-12-12/60-62/Mülhausen.

25 Schwarzschild, Karl: Über den Einfluß von Wind und Luftdichte auf die Flugbahn der Geschosse, 1920, p. 37–63.

interest of Karl Schwarzschild, who looked for trial functions of the metric coefficients by good luck, which produced a line-element in "curved spacetime", which fulfilled the conditions of general relativity and is singular at the origin only. The Schwarzschild-metric, slightly modified by David Hilbert (1862–1943), produces exact analytical solutions of the relativistic field equations in terms of elliptical integral-functions of Karl Weierstraß (1815–1897). Early 1916, Einstein has reported regularly from their correspondence about the Gravitational Field in *"curved spacetime"* to the *Prussian Physical Society* and subsequently supported the publication of Schwarzschild's outer[26] and inner[27] solutions of the gravitational fields caused by a non-rotating mass.

4.6 Literatur

CHANT, CLARENCE AUGUSTUS: The Mount Wilson Conference of the Solar Union. In: *Journal of the Royal Astronomical Society of Canada* **4** (1910), p. 356–372.

HERRMANN, DIETER B.: *Ejnar Hertzsprung – Pionier der Sternforschung.* Berlin, Heidelberg: Springer 2011.

HERTZSPRUNG, EJNAR: Zur Strahlung der Sterne. In: *Zeitschrift für Wissenschaftliche Photographie, Photophysik und Photochemie* **3** (1905), Heft 11, p. 429–442.

HERTZSPRUNG, EJNAR: Eine Annäherungsformel für die Abhängigkeit zwischen Beleuchtungshelligkeit und Unterschiedempfindsamkeit des Auges. In: *Zeitschrift für Wissenschaftliche Photographie, Photophysik und Photochemie* **5** (1907), Heft 12, p. 468–472.

HERTZSPRUNG, EJNAR: Über die räumliche Verteilung der Veränderlichen vom δ-Cephei-Typus (On the spatial distribution of variable [stars] of the δ-Cephei-Type). In: *Astronomische Nachrichten* **196** (1913), No. 4692, p. 201–208.

LEPSIUS, B. & R. WACHSMUTH: *Denkschrift der ersten internationalen Luftschiff-fahrts-Austellung (ILA) zu Frankfurt am Main, 1909. Band I – Wissenschaftliche Vorträge.* Berlin: Springer 1910, p. 107–108.

SCHWARZSCHILD, KARL & O. BIRCK: *Tafeln zur astronomischen Ortsbestimmung im Luftballon bei Nacht, sowie zur leichteren Bestimmung der mitteleuropäischen Zeit an jedem Orte Deutschlands: mit Unterstützung der Göttinger Vereinigung für angewandte Mathematik und Physik.* Göttingen: Vandenhoeck & Ruprecht 1909.

SCHWARZSCHILD, KARL: Über einen Transformator zur Auflösung sphärischer Dreiecke, besonders für Zwecke der Ortsbestimmung im Luftballon. In: *Zeitschrift für Instrumentenkunde* **30** (1910), p. 75–80.

26 Schwarzschild, Karl: Über das Gravitationsfeld eines Massenpunktes nach der Einsteinschen Theorie, 1916, p. 189–196.

27 Schwarzschild, Karl: Über das Gravitationsfeld einer Kugel aus inkompressibler Flüssigkeit nach der Einsteinschen Theorie, 1916, p. 424–434.

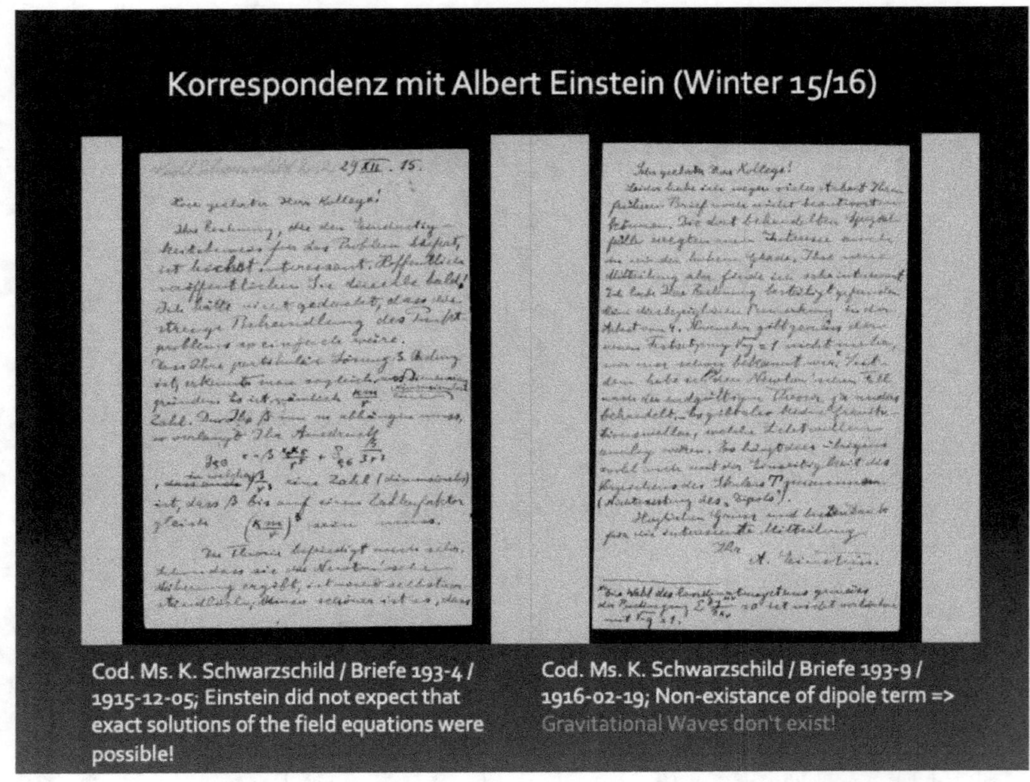

Figure 4.14:
Postcards of the correspondence between
Karl Schwarzschild and Albert Einstein, 1915/16

(Cod. Ms. K. Schwarzschild, Briefe 193-4 (links) und 193-9 (rechts): Einstein, Albert)
(Nachlass K. Schwarzschild; HANS SUB Universität Göttingen)

SCHWARZSCHILD, KARL: Künstlicher Horizont und Ballonsextant. In: *Zeitschrift für Instrumentenkunde* **30** (1910), p. 357–359.

SCHWARZSCHILD, KARL: Libellenhorizont und Libellensextant. In: *Zeitschrift für Flugtechnik und Motorluftschiffahrt* **4** (1913), p. 177–180 (4 Seiten).

SCHWARZSCHILD, KARL: Über den Einfluß von Wind und Luftdichte auf die Flugbahn der Geschosse (engl.: On the Influence of Wind and Density of Air on the Trajectory of Shells). In: *Sitzungsberichte der Preussischen Akademie der Wissenschaften Berlin* 1920, p. 37–63.

SCHWARZSCHILD, KARL: Über das Gravitationsfeld eines Massenpunktes nach der Einsteinschen Theorie (engl.: On the Gravitational Field of a Point Mass according to Einstein's Theory). In: *Sitzungsberichte der Preussischen Akademie der Wissenschaften Berlin* 1916, p. 189–196.

Schwarzschild, Karl: Über das Gravitationsfeld einer Kugel aus inkompressibler Flüssigkeit nach der Einsteinschen Theorie (engl.: On the Gravitational Field of a Sphere of incompressible Fluid according to Einstein's Theory). In: *Sitzungsberichte der Preussischen Akademie der Wissenschaften Berlin* 1916, p. 424–434.

Abbildung 5.1:
Hans Kienle (1895–1975), um 1950

Porträtsammlung des Deutschen Museums, München, Archiv (http://www.digiporta.net/index.php?id=265052790).

Hans Kienle (1895–1975) – Wissenschaftler zwischen Astrophysik und Politik im 20. Jahrhundert

Maik Schmerbauch (Berlin)

Abstract:

The life of the German astronomer Hans Kienle (1895–1975) is a quite a challenge for the history of astronomy. This is because his professional life in astronomy took place within the context of a wide variety of political systems in German history in the 20th century. He carried out many experiments in the field of astronomy and formed research results, which gave him a great reputation in Germany and internationally until the end of his life. Reconstructing his professional life, but also private events, is therefore a major but worthwhile task. The article aims to outline the life of the astronomer Kienle in its stages, also using newly researched sources in archives. The archives are named where the most important documents on the life and work of the astronomer Kinele are located. This is intended to lay the foundations for further extensive research on Hans Kienle, and thus also on the history of German astronomy in the 20th century.

Zusammenfassung

Das Leben des deutschen Astronom Hans Kienle (1895–1975) ist für die Astronomiegeschichte eine gute Herausforderung. Denn sein berufliches Leben für die Astronomie fand im Rahmen unterschiedlichster politischer Systeme der deutschen Geschichte im 20. Jahrhundert statt. Auf dem Gebiet der Astronomie hat er viele Untersuchungen durchgeführt und zu Forschungsergebnissen geformt, was ihm in Deutschland und international eine große Reputation bis zu seinem Lebensende bescherte. Die Rekonstruktion seines beruflichen Lebens, aber auch privater Ereignisse, ist daher eine große, aber lohnenswerte Aufgabe. Der Beitrag möchte das Leben des Astronomen Kienle in seinen Etappen skizzieren, und zwar auch anhand neu recherchierter Quellen in Archiven. Es werden die Archive benannt, in denen sich die wichtigsten Dokumente zum Leben und Wirken des Astronoms Kinele befinden. Damit sollen die Grundlagen für eine große weitergehende Forschung über Hans Kienle, und damit auch über die deutsche Astronomiegeschichte des 20. Jahrhunderts, gelegt werden.

5.1 Einleitung

Biographien über bekannte deutsche Astronomen gibt es bislang in Deutschland noch recht wenig. Über die in der Öffentlichkeit populären Astronomen und Astrophysiker wie Hermann Oberth (1894–1989), Karl Schwarzschild (1873–1916), Sigmund Jähn (1937–2019) und im weiteren Sinn auch Albert Einstein (1879–1955) oder Wernher von Braun (1912–1977) gibt es bereits wissenschaftliche Untersuchungen.[1]

Über bedeutende Astronomen, die das Fach an den Lehrstühlen deutschen Universitäten über Jahrzehnte in besonderem Maß geprägt haben, liegen bislang aber noch kaum Forschungen vor. Unter ganz vielen Namen berühmter Inhaber von deutschen Astronomielehrstühlen und Leiter der zugehörigen Sternwarten und Observatorien, können hier nur beispielsweise genannt werden Astronomen wie Hans Heinrich Voigt (1921–2017), Carl Wilhelm Wirtz (1876–1939), Hans-Ludwig Neumann (1938–1991), Friedrich Wilhelm Hans Ludendorff (1873–1941), Friedrich Simon Archenhold (1861–1939), Otto Hachenberg (1911–2001), Hugo von Seeliger (1849–1924) oder auch Harry O. Ruppe (1929–2016). In Diensten der „Astronomischen Gesellschaft“ ließen sich noch zahlreiche andere Namen finden. Auch Frauen waren immer wieder der Astronomie eng verbunden. Als ein Name, wenn auch nicht in akademischen Rängen, ist hier Alice Archenhold (1874–1943)[2] als ein Beispiel zu nennen, die dem Rassenwahn der Nazis zum Opfer fiel.

Selbstverständlich ist aus geschichts- und archivwissenschaftlicher Sicht eine Bearbeitung von Biographien von Astronomen des 20. Jahrhundert die wichtigste Grundlage, zunächst die Quellen zu identifizieren, die aber aufgrund der

1 Vgl. Barth: Hermann Oberth, 2008. Reinsch & Wittmann: Karl Schwarzschild, 2017. Hoffmann: Sigmund Jähn, 2008.

2 Lübcke: Alice Archenhold (1874–1943), 2002, S. 13–20.

jungen Laufzeiten oft noch gar nicht dem zuständigen Archiv vollständig bekannt oder auch noch nicht erschlossen worden sind. Auch liegen auf wichtigen Archivalien gerade für die Zeit von 1945 bis 1990 gelegentlich auch noch Schutz- und Sperrfristen, so dass ein Zugang zu diesen nicht ohne Weiteres und immer einfach möglich ist. Hinzu kommen in neuester Zeit auch verschiedenste Datenschutzbestimmungen, die den Zugang zu Archivalien nicht einfacher machen werden. Viele wissenschaftliche Biographien aus der Geschichtswissenschaft konzentrieren sich auch Sicht des Archivars deshalb erkennbar zu sehr auf einzelne Überlieferungsarchive, manchmal auch geschuldet den zeitlichen Gründen der Fertigstellung einer wissenschaftlichen Arbeit, öfter aber aufgrund der mangelnden quellenkundlichen Kenntnis. Denn eine fundierte Biographie, die das Leben eines Menschen erfüllend und auch aus verschiedenen Perspektiven beschreiben soll, muss möglichst viele Provenienzen abdecken, und auch die Zeit dafür den Forschenden vorhanden sein. So ist eine Biographie eines Astronomen aus dem 20. Jahrhundert auch eine Herausforderung für einen Astronomiehistoriker und deshalb ein Langzeitprojekt. Für eine Biographie des deutschen Astronomen Hans Kienle (1895–1975) hat der Autor bereits eine Menge an archivischen Vorarbeiten getan, die eine Beschäftigung mit dem Leben dieses Astronomen in den nächsten Jahren sicher ermöglichen.

Über Hans Kienle ist bislang wenig wissenschaftlich geforscht worden. Ein Artikel seines Assistenten Otto Heckmann (1901–1983) seiner Göttinger Zeit in Form eines kurzen Nachrufs erschien kurz nach Kienles Tod.[3] Einen kurzen und komprimierten Beitrag publizierte Kummer (1996) in der Zeitschrift *„Sterne und Weltraum“*. Ein weiterer Artikel erschien im Jahr 2000 von Günther Wirth.[4]

In anderen historischen Abhandlungen findet sich sein Name oft auch in anderen Zusammenhängen mit der Astronomie und Wissenschaftsgeschichte. Er selbst hat während seiner Tätigkeit hunderte fachwissenschaftliche Artikel publiziert. Allein deren Aufzählung würde den Rahmen des Beitrags sprengen. So ist Hans Kienle für die Astronomiegeschichte neben diesen astrophysikalischen Forschungsergebnissen vor allem interessant, da er seinen Beruf unter den verschiedensten politischen Systemen der neuesten deutschen Geschichte im 20. Jahrhundert ausüben musste. Geboren 1895 im oberbayrischen Kulmbach und aufgewachsen im Zweiten Deutschen Kaiserreich, begann und beendete er noch während des Ersten Weltkrieges sein astronomisches Studium, machte dann eine steile Karriere in der Weimarer Zeit, erlebte den Nationalsozialismus von Anfang bis zum Untergang, sowie die anschließende Besatzungs-

3 Heckmann: Nachruf auf Hans Kienle, (1976), S. 9–11.

4 Wirth: Weltanschauliche und wissenschaftstheoretische Aspekte im Werk Hans Kienles, 2000, S. 151–168.

und Nachkriegszeit in der Sowjetisch-Besetzten Zone, und forschte dann noch einige Jahrzehnte in der BRD, wo er 1975 starb.

Dass er eine bedeutende Person der akademischen Astronomie in Deutschland war, und darüber hinaus international gut vernetzt, zeigen seine zahlreichen Publikationen, Ämter und Reputationen in Deutschland und im Ausland. Auch wurde er in Standardwerken der Astronomie und in Lexika erwähnt. Als ein Beispiel findet sich im mehrteiligen *„Lexikon der Astronomie"* eine Kurzbiographie über ihn, und zwar aufgrund seiner *„wichtigen Arbeiten über Sterntemperaturen, Astrospektroskopie und Kosmogonie".*[5] Im *„Brockhaus der Astronomie"* von 2005 ist Kienle ebenfalls vertreten,[6] und auch im *„International Who is Who"* fand er regelmäßig seinen Platz.[7] Für das Entwerfen einer Biographie über Hans Kienle ist aber die Kenntnis der archivalischen Quellen die unbedingte Grundlage, auf der die wissenschaftlichen Forschungen dann sukzessive stattfinden können. Dieser Beitrag will einen Überblick über die verschiedenen Lebensstationen Kienles geben, und auch die archivische Situation in den Blick nehmen.

5.2 Kindheit und Jugend

Die Kindheit und Jugend von Hans Kienle könnten sich aus familiären und privaten Quellen als auch aus seinen späteren fachwissenschaftlichen Korrespondenzen, Berichten, Erklärungen und Erinnerungen rückblickend rekonstruieren lassen. Auch andere Personen, die ihn aus dieser Zeit kannten oder Zeugnisse über ihn hinterlassen haben, könnten Informationen zuliefern. Der Zugriff auf diese privaten Informationen aber setzen aber einen sehr großen Rechercheaufwand voraus, der auch nicht zu sehr in die Tiefe gehen sollte, da seine berufliche Seite weitaus ausschlaggebender für eine Biographie im Bereich der Astronomie ist als seine Kindheit. Für die Kindheit Kienles sind auch die frühen Immatrikulationsunterlagen in München ab 1915 interessant. Einiges an Vorarbeit geleistet hat auch Kummer (1996) mit seinem kurzen Beitrag zur Kindheit Kienles.

Hans Kienle, eigentlich Johann Georg Kienle, wurde geboren am 22. Oktober 1895 in Kulmbach. Sein Vater war Lucien Kienle, von Konfession evangelisch-lutherisch, und stammte aus dem Raum Nürnberg.[8] Von Beruf war der Vater

5 Elsässer: Lexikon der Astronomie, 1995, Artikel Hans Kienle, S. 339.

6 Kilian & Sauermost (Red.): Brockhaus der Astronomie, 2005; Artikel Hans Kienle, S. 218.

7 International Who is Who, 1968, Artikel Kienle, Johann (Hans) Georg.

8 Universitätsarchiv Heidelberg, Personalakte Hans Kienle, Signatur PA 4481; Personalbogen vom 14. Oktober 1950, 4. S.

Gürtler gewesen. Seine Mutter war die Katharina, und zwar „in einfache Verhältnisse“.[9] Kienle besuchte die Volksschule in Nürnberg von September 1901 bis Juli 1905, wechselte anschließend auf die Realschule Nürnberg, die er im Juli 1911 verließ. Ab September 1911 besuchte er bis Juli 1914 die Oberrealschule in Nürnberg und legte die Matura ab.[10]

Schon zu seiner Schulzeit prägten sich spätere Charakterzüge des Wissenschaftlers Kienle heraus. Er scheute nicht den Konflikt mit den Lehrern, hinterfragte Inhalte kritisch, was naturwissenschaftliche Fragen, aber auch religiöse Fragen betraf. In einem Fall hat er sogar sensible Unterrichtsinhalte an das Bayerische Schulministerium weitergeleitet, und dafür einen Verweis kassiert. Dennoch gab es auch viele Lehrer, die er respektierte und die ihn prägten. In Kienles Reifezeugnis wurde im dann attestiert *„Sein Fleiß war immer hervorragend groß, sein Verhalten gab Proben einer idealen Geistesrichtung“*.[11] Kienle beendete seine Schulausbildung im Sommer 1914. Doch auf ihn wartete wie viele andere junge Millionen Deutsche nun eine ganz andere Situation des beruflichen Werdegangs: der Erste Weltkrieg brach im August 1914 aus, und dessen anfängliche Begeisterung erfasste den jungen Schulabgänger Kienle.

5.3 Ein Studium während des Ersten Weltkrieges 1915–1918

Für seine Zeit des Studiums an der Ludwig-Maximilians-Universität (LMU) München und seine Assistentenzeit an der Sternwarte können seine Immatrikulationsakten als auch die Unterlagen seines Studiums und seiner späteren Anstellung herangezogen werden, die im Universitätsarchiv der LMU zugänglich sind. Auch finden sich zu seiner kurzen Beschäftigungszeit als Landesbediensteter wenige Dokumente im Bayerischen Staatsarchiv und auch im Archiv des Deutschen Museums München. Damit lassen sich seine beruflichen Jahre 1915 bis 1924 grundsätzlich, aber zeitaufwendig, rekonstruieren.

Einige der wenigen Autoren, die über Kienle posthum schrieben, erwähnen, Kienle hätte nach seinem Schulabschluss ein Stipendium von 150 RM bekommen, so dass er sich schon um Ostern 1914 an der LMU in Astronomie immatrikulierte.[12] Seine Immatrikulationskartei der LMU aber zeigt keinen Hinweis, dass er schon zum Sommersemester 1914 ein Studium begonnen hatte, sondern

9 Kummer: Hans Kienle, (1996), S. 266–269.

10 Universitätsarchiv Heidelberg, Personalakte Hans Kienle, Signatur PA 4481; Personalbogen vom 14. Oktober 1950, 4. S.

11 Kummer: Hans Kienle, (1996), S. 266–269.

12 Kummer: Hans Kienle, (1996), S. 266–269.

erst im Januar 1915.[13] Den Studienbeginn zu diesem Zeitpunkt 1915 belegen auch andere Quellen,[14] auch wenn Kienle am 17. November 1920 dem Rektorat der Universität mitteilte, er habe am 27. Mai 1914 die Aufnahmeurkunde der Universität bekommen, obwohl er zu diesem Zeitpunkt noch in der Schule war.[15] Fakt ist, Kienle war erst 19 Jahre alt, als der Erste Weltkrieg im August 1914 ausbrach. Auch er meldete sich als Kriegsfreiwilliger, und wurde noch im August 1914 eingezogen. Er erlitt aber eine schwere Kriegsverletzung an der Westfront, er verlor nämlich das rechte Auge, und wurde nach eigenen Angaben im Dezember 1914 wieder aus der Kaiserlichen Armee entlassen.[16] Für seinen Einsatz erhielt er das Eiserne Kreuz II. Klasse verliehen.[17]

Kienle schien sich von der schweren Kriegsverletzung und den traumatischen Kriegserlebnissen als junger Mann darin nicht beeinflussen zu lassen, sich am 11. Januar 1915 an der LMU in Mathematik und Astronomie zu immatrikulieren.[18] Die Gründe, warum sich Kienle für das Studium der Astronomie entschieden hatte, sind nicht ganz geklärt. Eine Quelle viele Jahrzehnte später, als er schon in seinem Ruhestand war, berichtet, dass Kienle erzählt hatte. für das Studium hätte ihm *„das Erscheinen des Halleyschen Kometen 1910 wesentlichen Anlaß gegeben."*[19] Der Verlust eines Auges für ein Fach, das durch und von Beobachtung lebt, war für Kienle also kein Grund zur Resignation. Als Abschluss gab es nur die Anfertigung einer Dissertation. Kienle blieb während des gesamten Ersten Weltkrieges immatrikulierter Student. Er wäre damit auch ein guter Zeuge der Situation der bayerischen Großstadt während des Krieges gewesen, hat aber über diese Zeit nichts bislang bekanntes Privates hinterlassen.[20] Schon im Oktober 1915 bekam er eine Stelle als Hilfsassistent an der Sternwarte München.[21] Sein Examen machte er am 22. April 1918 mit dem

13 Universitätsarchiv der Ludwig-Maximilian-Universität München, Studienkartei Hans Kienle (A5).

14 Bundesarchiv Berlin, Bestand R 4901/13268, Reichsministerium für Erziehung, Wissenschaft und Forschung, Karteikarte zu Hans Kienle, Sig. K142.

15 Universitätsarchiv der Ludwig-Maximilian-Universität München, Akte E II 1980, Personalakte, Kienle an Rektorat am 21. November 1920, hss. Dok., 1 S.; Antwort auf das Schreiben des Rektorats vom 17. November 1920.

16 Universitätsarchiv Heidelberg, Personalakte Hans Kienle, Signatur PA 4481; Personalbogen vom 14. Oktober 1950, 4. S.

17 Bundesarchiv Berlin, Bestand R 4901/13268, Reichsministerium für Erziehung, Wissenschaft und Forschung, Karteikarte zu Hans Kienle, Sig. K142.

18 Universitätsarchiv der Ludwig-Maximilian-Universität München, Studienkartei Hans Kienle (A5).

19 Frankfurter Allgemeine Zeitung vom 27. Oktober 1965, Artikel Hans Kienle 70 Jahre.

20 Universitätsarchiv der Ludwig-Maximilian-Universität München, Studienkartei Hans Kienle (A5).

21 Universitätsarchiv Heidelberg, Personalakte Hans Kienle, Signatur PA 4481; Personalbogen vom 14. Oktober 1950, 4. S.

Dr. phil.[22] Am 1. März 1918 wurde er zum angestellten Assistenten an der Sternwarte München berufen.[23]

In dieser Zeit, in der auch der Erste Weltkrieg mit allen seinen politischen Folgen in München endete,[24] fertigte er auch seine Habilitationsschrift mit dem Thema *„Untersuchungen über Saalrefraktion“*. Am Mittwoch, den 21. Juli 1920, fand die Probelesung zur Erlangung der Venia-Legendi in der kleinen Aula der Philosophischen Fakultät unter dem Vorsitz von Prof. Arnold Sommerfeld (1868–1951) statt. Das Thema der Probe-Vorlesung hieß *„Zeitmessung und Zeitmeßapparate in der Astronomie“*.[25] Am 21. Juli schrieb Sommerfeld an den Senat, Kienle habe das Verfahren gut bestanden und man sah ihn als Dozenten ab dem Wintersemester vor.[26]

So arbeitete er drei Jahre an der Münchner Sternwarte und der LMU, forschte und publizierte, so dass er über München hinaus schnell bekannt wurde. Am 3. November 1923 bekam Kienle dann Post vom Preußischen Ministerium für Wissenschaft, Kunst und Volksbildung aus Berlin ob er bereit wäre, den Professor für Astronomie Johannes Hartmann (1865–1936) an der Universität Göttingen zu vertreten. Kienle hatte sich dazu *„prinzipiell bereit erklärt“*, diese berufliche Chance zu übernehmen.[27] Genau zu diesem Zeitpunkt war die politische Stimmung in München äußerst aufgeheizt, denn so stand auch der Hitler-Putsch kurz bevor. Ob auch diese politischen Ereignisse ihn in die niedersächsische Provinz nach Göttingen bewegten, ist zumindest nicht ganz auszuschließen. Tatsache aber war, er bekam als einer der wenigen Wissenschaftler seiner Zunft eine Professur der Astronomie angeboten, ein Karriereschritt, der in dieser Zeit wohl nur wenigen Astronomen vergönnt war.

22 Universitätsarchiv der Ludwig-Maximilian-Universität München, Studienkartei Hans Kienle (A5).

23 Universitätsarchiv der Ludwig-Maximilian-Universität München, Akte E II 1980, Personalakte, S. 4.

24 Reif: Hunger, Wohnungsnot, Arbeitslosigkeit, 1995.

25 Universitätsarchiv der Ludwig-Maximilian-Universität München, Akte E II 1980, Personalakte, Einladung zur Probevorlesung von Hans Kienle am 21. Juli 1920.

26 Universitätsarchiv der Ludwig-Maximilian-Universität München, Akte E II 1980, Personalakte, Sommerfeld an den Akademischen Senat am 21. Juli 1920, hss. Dokument, 1 S.

27 Universitätsarchiv der Ludwig-Maximilian-Universität München, Akte E II 1980, Personalakte, Kienle an Dekan der Philosophischen Fakultät II der LMU am 3. November 1923, mss. Dok., 1 S., Abschrift.

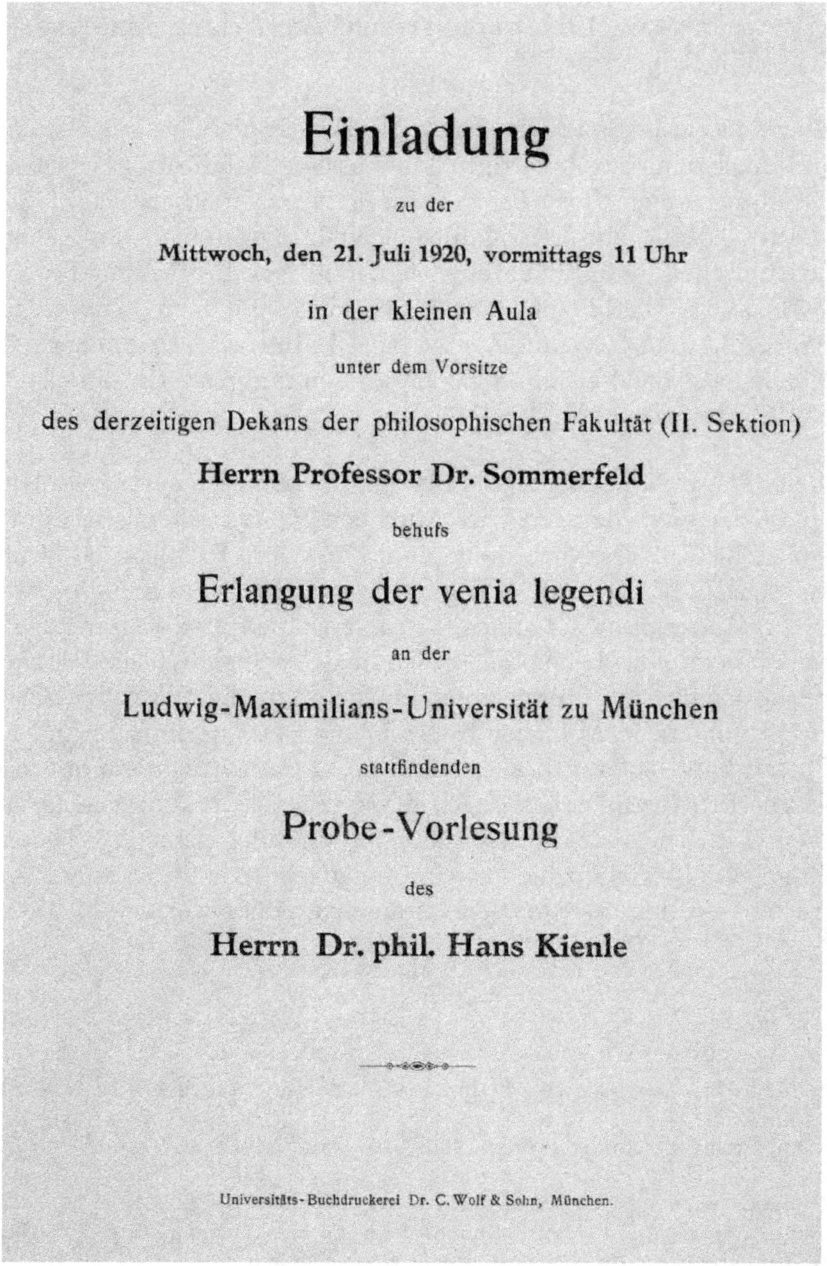

Einladung

zu der

Mittwoch, den 21. Juli 1920, vormittags 11 Uhr

in der kleinen Aula

unter dem Vorsitze

des derzeitigen Dekans der philosophischen Fakultät (II. Sektion)

Herrn Professor Dr. Sommerfeld

behufs

Erlangung der venia legendi

an der

Ludwig-Maximilians-Universität zu München

stattfindenden

Probe-Vorlesung

des

Herrn Dr. phil. Hans Kienle

Universitäts-Buchdruckerei Dr. C. Wolf & Sohn, München.

Abbildung 5.2:
Habilitation – Einladung zur Probevorlesung von Hans Kienle, 21.7.1920

(Universitätsarchiv der Ludwig-Maximilian-Universität München, Akte E II 1980, Personalakte, Einladung zur Probevorlesung von Hans Kienle am 21. Juli 1920)

5.4 Wechsel an die Universität nach Göttingen

Für die Göttinger Zeit können für die ersten zwei Jahre noch wenige Quellen aus dem Archiv der LMU zuliefern, aber auch aus Archiv des Deutschen Museums. Die Hauptprovenienz aber ist das Universitätsarchiv Göttingen, das noch umfangreich ausgewertet werden muss. Auch die Sammlung der frei verfügbaren und digitalisierten Vorlesungsverzeichnisse ist eine Fundgrube zur akademischen Aktivität Kienles an der Georg-August-Universität von 1924 bis 1939. Die Universität Göttingen war schon seit der Mitte des 18. Jahrhunderts zu einem bedeutenden Ort der astronomischen Forschung geworden. Eine eingerichtete Sternwarte lieferte seit dieser Zeit bedeutende astronomische Kenntnisse. Auch die anderen Naturwissenschaften wie Mathematik und Physik waren etablierte Fächer in Göttingen.[28] Insofern bot sie für den jungen Kienle einen guten Karriereschritt in die freie astronomische Forschung, konnte er seinen Schwerpunkten auf einem eigenen Lehrstuhl nun angehen, wenn auch zunächst vertretungsweise. Der Preußische Minister für Wissenschaft, Kunst und Volksbildung bestellte ihn zum 1. Mai 1924 als außerplanmäßigen Professor für Astronomie. Seine Besoldung von 3936 Goldmark kurz nach der Inflation war eine stattliche Entlohnung für den jungen Wissenschaftler. Auch sicherte die Georg-August Universität Kienle die Möglichkeit akademischer Lehre zu, wenn Professor Hartmann wieder aus dem Ausland zurückkommen werde.[29] Kienle nahm die Lehre zum Wintersemester 1924 auf, und zwar an der Mathematisch-Naturwissenschaftlichen Fakultät. Er wurde dort als *„nichtbeamteter außerplanmäßiger Professor“* für Astronomie geführt.[30] Seine ersten Veranstaltungen 1924 waren eine einführende Vorlesung *„Grundlagen der Astronomie“* und Unterricht zur „Himmelsmechanik“.[31]

Wohl eher unerwartet traf Kienle schon nach wenigen Monaten auch sein privates Glück an der Universität. Dort lernte er die wissenschaftliche Assistentin der Zoologie kennen, Elsa Maria Armbruster, die auch einige Jahre älter als er war. Die Beziehung wurde schnell innig, denn schon am 27. Dezember 1924 heiratete Kienle seine Elsa zu seiner Frau.[32] Wie wichtig dem Astronomen trotz

28 Geyken: Zum Wohle Aller. Geschichte der Georg-August-Universität Göttingen, 2019.

29 Universitätsarchiv Heidelberg, Personalakte Hans Kienle, Signatur PA 4481. Der Preußische Minister für Wissenschaft, Kunst und Volksbildung an Kienle am 11.4.1924, mss. Dok., 1 S.

30 Vorlesungsverzeichnis der Georg-August Universität zu Göttingen für das Wintersemester 1924. Göttingen 1924, S. 9 f.

31 Vorlesungsverzeichnis der Georg-August Universität zu Göttingen für das Wintersemester 1924, S. 21.

32 Bundesarchiv Berlin, Bestand R 4901/13268, Reichsministerium für Erziehung, Wissenschaft und Forschung, Karteikarte zu Hans Kienle, Sig. K142.

dieser eindrucksvollen Karriere, immer und jederzeit, auch sein privates Glück war, zeigte sich alsbald, als aus dieser Ehe ein reicher Nachwuchs hervorging. So wurde Kienle in seiner Göttinger Zeit der Vater von drei Töchtern: Johanna Margarete (geb. 13.12.1925), Marie Luise (22.2.1928) und Margarete Ursula (*15.2.1932).[33]

Kienle bekam auch finanzielle Mittel in Göttingen, um für sich Hilfskräfte anzustellen. Eine Anzeige startete er im April 1926 in den *„Astronomischen Nachrichten"* mit der schlichten Angabe, es sei von ihm ein *„Jüngerer Astronom mit guter physikalischer Durchbildung als wissenschaftlicher Hilfsarbeiter gesucht. Bewerbungen an H. Kienle, Sternwarte Göttingen"*.[34]

Auch konnte er verschiedene Expeditionen in das Ausland durchführen, wo er seine astronomischen Beobachtungen machte. 1927 beobachtete er zum Beispiel eine Sonnenfinsternis in Schweden in Gällivare, zu der er mit einigen Kollegen und seiner Frau aufgebrochen war.[35]

Im gleichen Jahr (1927) bot man ihm auch die feste Professur für Astronomie an. Am 18. Juli 1930 wurde Kienle dann ein ordentliches Mitglied der *Gesellschaft der Wissenschaften zu Göttingen*, und er wurde fest in die akademische Selbstverwaltung integriert.[36] Doch schon kurze Zeit später wurde er mit einer neuen politischen Gefahr konfrontiert, als die Nationalsozialisten am 30. Januar 1933 die Macht in Deutschland übernahmen. Auch er und die Astronomie hatten in der folgenden Zeit des Dritten Reiches unter diesem politischen System Druck und Leid erfahren.

5.5 Die Machtübernahme der Nationalsozialisten

Hans Kienle gehörte zu den Astronomie-Professoren in Deutschland, die über die gesamte Zeit des Dritten Reiches 1933–1945 im Wissenschaftssystem des NS-Staates integriert waren. Weitere Astronomen mit ähnlicher Karriere in der NS-Zeit waren z. B. der Astronom Paul Guthnick (1879–1947), der aber noch einer anderen gesonderten Forschung bedarf. Kienle blieb bis 1939 an der Universität Göttingen und wechselte dann an die Universität Berlin und übernahm als Direktor auch das Astrophysikalische Observatorium in Potsdam.

Über Kienles Zeit im Dritten Reich ist eine eigene große Abhandlung notwendig, so dass hier nur einige Grundzüge aus diesen zwölf Jahren des Terrors

33 Universitätsarchiv Heidelberg, Personalakte Hans Kienle, Signatur PA 4481; Personalbogen vom 14. Oktober 1950, 4. S.

34 Kienle (1926), S. 271.

35 Kummer: Hans Kienle, (1996), S. 266–269.

36 Bundesarchiv Berlin, Bestand R 4901/13268, Reichsministerium für Erziehung, Wissenschaft und Forschung, Karteikarte zu Hans Kienle, Sig. K142.

und Krieges vorgestellt werden können. Die Quellenlage für diesen Zeitraum ist sehr umfangreich. So ist für die Jahre 1933 bis 1945 weiter das Universitätsarchiv Göttingen wichtig. Als weitere Archive sind unbedingt zu sichten das Universitätsarchiv der Humboldt-Universität Berlin, das Archiv der Deutschen Akademie der Naturforscher Leopoldina, das Bundesarchiv Berlin, und auch das Archiv des Deutschen Museums. In diesen lagern hunderte Seiten an schriftlichen Dokumenten, die ausgewertet werden müssen.

Die Nationalsozialisten erließen im April 1933 das *„Gesetz zur Wiederherstellung des Berufsbeamtentums“*, das nichts anderes festlegte, als das „nichtdeutsche“, und im NS-Jargon bezeichnete „nichtarische“ Beamte, aus allen staatlichen Behörden zu entlassen waren. Es wurde aber in den Folgejahren auch zur „Entfernung“ aller deutschen Beamten genutzt, die sich nicht klar für den Nationalsozialismus positionierten. Gerade die Universitäten mit ihrem großen Einfluss auf die jungen Studenten und auf die gesamte Forschung und Bildung im Deutschen Reich wurden von den Nationalsozialisten personell „gesäubert“. So wurden auch jüdische Astronomen in Deutschland, an ihrer bislang sehr freien Tätigkeit plötzlich behindert, verfolgt, und oft auch aus Deutschland vertrieben worden.[37] Als nur beispielhafte Opfer jüdischer Abstammung unter den deutschen Astronomen seien an dieser Stelle Friedrich Simon Archenhold (1861–1939),[38] sein Sohn Günter Archenhold (1904–1999)[39] oder auch Kienles Promovend Martin Schwarzschild (1912–1997)[40] genannt. Das grausame Schicksal der Frau von Friedrich Simon Archenhold und der Mutter von Günter Archenhold, Alice Archenhold (1874–1943), die ihren Mann in vielen Fragen der Astronomie unterstützte, bestürzt die Astronomiegeschichte bis heute. Sie wurden von den Nationalsozialisten deportiert und starb wie ihre Tochter Hilde im Konzentrationslager Theresienstadt 1943.[41]

Es soll an dieser Stelle Kienle selbst zur Sprache kommen, der seine Zeit im Nationalsozialismus im Januar 1946 gegenüber einer Entnazifizierungskommission rückblickend folgend dargestellt hat :[42]

> *„Weder ich noch meine Frau haben zu irgendeiner Zeit der NSDAP oder einer ihrer Gliederungen als Mitglieder angehört. Ich habe lediglich die „freien Beiträge“ zur NSV entrichtet, ohne der*

37 Pedde, Friedhelm: Verfolgung und Exil. Verfemte deutsche Astronomen im Dritten Reich. Teil 3. Martin Schwarzschild, Friedrich und Günter Archenhold, 2021, 8–11.

38 Herrmann: Friedrich Simon Archenhold und seine Treptower Sternwarte (1986).

39 Strach: Obituary Gunter Archenhold, (1999), S. 226.

40 Mestel: Martin Schwarzschild (1999), S. 469–484.

41 Lübcke: Alice Archenhold, 2002, S. 13–20.

42 Universitätsarchiv Heidelberg, Personalakte Hans Kienle, Signatur PA 4481; Bericht von Hans Kienle über seine Zeit während des Dritten Reiches, mss. Dok., 2. S.

Ehre der Mitgliedschaft teilhaftig gewesen zu sein. Ich habe den alten Luftsport-Verband im Interesse am Segelflug durch Beiträge gefördert, bin aber bei dessen zwangsweiser Eingliederung in das NSFK nicht aktives Mitglied dieser Organisation geworden. Den akademischen Organisationen (Dozentenschaft, NS-Dozentenbund) bin ich ferngeblieben und habe ebenso bei Übernahme der Leitung des Astrophysikalischen Observatoriums in Potsdam den Eintritt in die „zuständige Berufsorganisation", den „Reichsbund Deutscher Beamten", abgelehnt. Im Kriege habe ich einzelne, in mein spezielles Fachgebiet fallende Forschungsaufträge übernommen, um die wissenschaftliche Weiterarbeit des Observatoriums zu ermöglichen; massgebend bin ich an keiner der unmittelbaren Kriegsaufgaben beteiligt gewesen. Ab 1943 wurde ich auf Veranlassung von Staatsrat Esau als dessen Sachberater für Astronomie im Reichsforschungsrat geführt und wurde in dieser Eigenschaft in die letzte Phase der Ausplünderung der sowjetischen Sternwarten verwickelt. Es gelang mir, wenigstens die Bibliothek und das Plattenarchiv, sowie einige Reste der Instrumente der gegen meinen Einspruch unter dem Motto „Bergung wissenschaftlichen Gutes" abgebauten Sternwarte Simeis (Krim) zu retten und sicherzustellen, so daß sie jetzt dem rechtmäßigen Eigentümer zurückgegeben werden konnten. Bei der von einer Kommission der Akademie der Wissenschaften der UdSSR durchgeführten Untersuchung ist die Korrektheit meines Verhaltens festgestellt worden, so daß ich in keiner Weise etwa als Kriegsverbrecher belastet bin. Der Auszug aus Ihren Fakultätsakten, die Sie Ihrem Schreiben beilegten, bestätigt, daß ich damals als „politisch nicht zuverlässig" im Sinne der NSDAP galt. Als besonders diskriminierend wurde die akademische Rede erachtet, die ich bei der letzten Verfassungsfeier der Universität im Juni 1932 gehalten hatte. Bei den Vorstandswahlen zur Astronomischen Gesellschaft gelegentlich des internationalen Kongresses (August 33), den ich nach Göttingen eingeladen hatte, kam es zu einer Stichwahl zwischen [Heinrich] *Vogt* [1890–1968] *und mir; Vogts Wahl wurde durch die Anhänger der Partei durchgedrückt gegen die Stimmen der in zu geringer Anzahl anwesenden Ausländer.*

1935 wurde meine Berufung nach München abgelehnt mit der Begründung, daß ich als „Judenfreund" für die Hauptstadt der Bewegung untragbar sei; ebenso 1933 die Berufung nach Wien mit der Begründung, die großen Neubauten der Zukunft – die Wiener Sternwarte sollte verlegt werden – müssten „jungen tatkräftigen Natio-

> *nalsozialisten" vorbehalten bleiben (Thüring!). Stattdessen bot man mir Potsdam an mit der zweifelhaften Schmeichelei:* „Sie haben in Göttingen gezeigt, was man mit bescheidenen Hilfsmitteln machen kann; wir erwarten von Ihnen daß Sie auch mit den etwas veralteten Potsdamer Einrichtungen tüchtige Arbeit leisten werden." *Als Kompensation für die mangelnden Verdienste um die Partei hat man mir bei dieser Berufung zugute gehalten, daß ich 1914 als Kriegsfreiwilliger mitgekämpft habe. Durch die Übernahme des Postens eines Direktors des Astrophysikalischen Observatoriums in Potsdam ging ich meiner Rechte als Ordinarius (seit 1927) verlustig, da die gleichzeitige Ernennung zum persönlichen Ordinarius an der Berliner Universität verweigert wurde. Das nahm ich allerdings nicht tragisch damals, da ich noch 20 Jahre vor mir hatte, bis die Frage „Emeritierung" oder „Pensionierung" akut geworden wäre. Wie ich jetzt erst erfuhr, ist auch meiner Wahl zum Mitglied der Preußischen Akademie der Wissenschaften im Jahre 1942 die Bestätigung durch das Ministerium versagt worden. Als letztes Glied fügt sich in die Kette der Diskriminierungen eine anonyme Anklage ein, die im Jahre 1943 unmittelbar bei der Reichskanzlei eingebracht wurde mit dem Antrag auf Enthebung vom Amt, da ich nicht Parteigenosse sei und* „mir alle Charaktereigenschaften fehlen, die die vornehme Führernatur auszeichen." *Diese Anklage hatte keine weiteren Folgen, da Herr v. Rottenburg, der mit ihrer Verfolgung beauftragt war, sie nach einer kurzen Aussprache mit einer Aktennotiz von mir in seiner Schieblade verschwinden ließ. Ich schreibe Ihnen das alles etwa nicht, um mich auch als ein „Opfer des Faschismus" hinzustellen, sondern nur, um Ihnen alle Unterlagen zu geben, die Sie bei einer Überprüfung meiner Personalien vielleicht benötigen. Ich muss es ganz Ihnen überlassen, ob uns welchen Gebrauch Sie davon machen wollen."*

Eine ausführliche Untersuchung über Kienles Zeit im Dritten Reiches wird viel an Wissen über seine Tätigkeit und der Situation der Astronomie ans Tagesicht bringen. Eine Abfrage an das frühere Berlin-Document-Center, des heutigen Bundesarchivs, bestätigte bereits einige Angaben Kienles als wahr. Denn das *„Reichsministerium für Erziehung, Wissenschaft und Kunst"* führte über alle Professoren im Deutschen Reich eine eigene Kartei, die auch Einträge für „Mitgliedschaften in nationalen Verbänden" dokumentierte. Für Kienle gab es dort keine Einträge hinsichtlich der Mitgliedschaft in NS-Verbänden oder auch

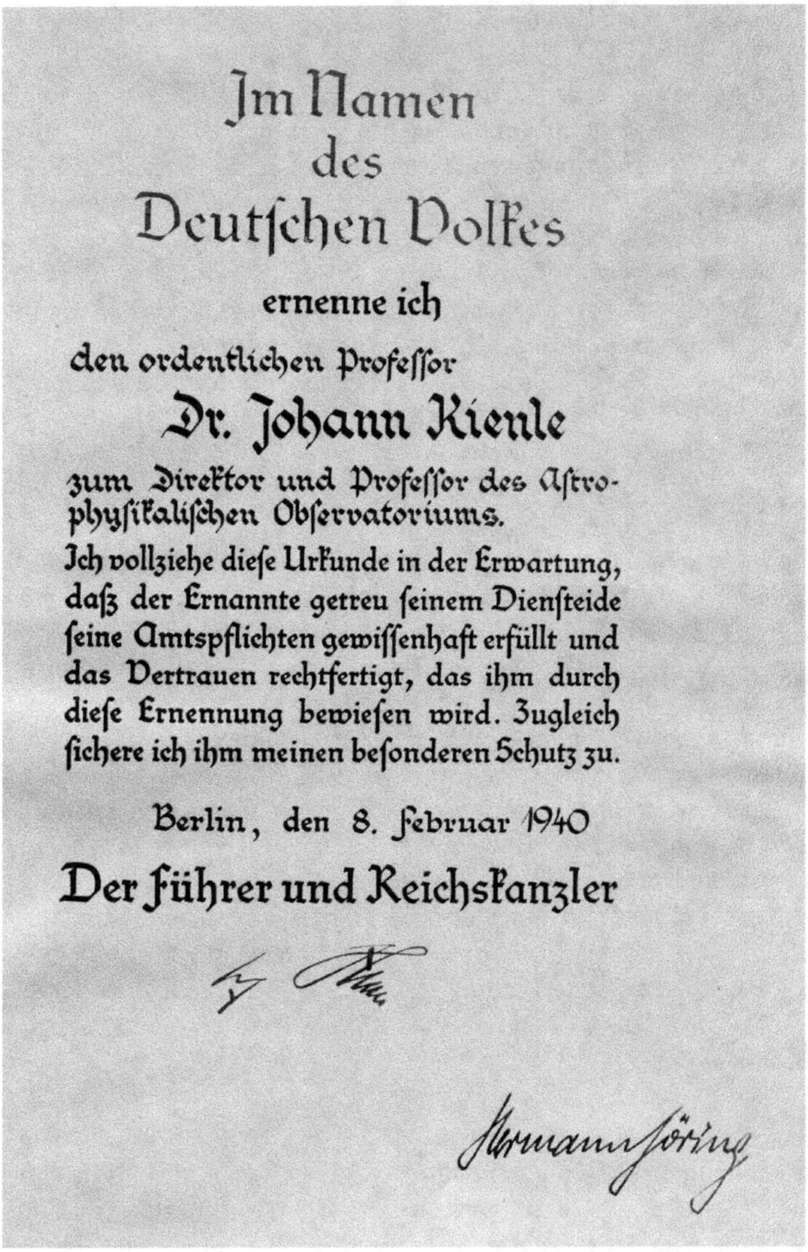

Im Namen
des
Deutschen Volkes

ernenne ich

den ordentlichen Professor

Dr. Johann Kienle

zum Direktor und Professor des Astrophysikalischen Observatoriums.

Ich vollziehe diese Urkunde in der Erwartung, daß der Ernannte getreu seinem Diensteide seine Amtspflichten gewissenhaft erfüllt und das Vertrauen rechtfertigt, das ihm durch diese Ernennung bewiesen wird. Zugleich sichere ich ihm meinen besonderen Schutz zu.

Berlin, den 8. Februar 1940

Der Führer und Reichskanzler

Abbildung 5.3:
Ernennungsurkunde Hans Kienles zum Direktor
des Astrophysikalischen Observatoriums Potsdam vom 8. Februar 1940

(Universitätsarchiv der Ruprecht-Karls-Universität Heidelberg,
Personalakte Hans Kienle, Signatur PA 4481)

der NSDAP.[43] Kienle sah sich aber keineswegs als ein „Opfer des Faschismus“, wusste er selbst genug über das Schicksal verfolgter jüdischer Wissenschaftler, darunter auch das Exil seines Doktoranden Martin Schwarzschild. Ein „Nichtwissen“ über die „Säuberungen“ und Verfolgungen von Wissenschaftlern an den Universitäten, wäre in seinen einflussreichen Positionen in Göttingen, Berlin und Potsdam in der Zeit 1933 bis 1945 und für einen Mann seines intellektuellen Profils für die Alliierten eher unglaublich gewesen, das muss er auch selbst eingesehen haben. Sein Ehrgeiz und seine Karriere sowie die Sorge um seine Familie erlaubte ihm aber keinen aktiven Widerstand gegen die Nationalsozialisten.[44]

Das Ende des Krieges erlebte er in Potsdam. Das Observatorium wurde am 14. April 1945 durch sowjetischen Beschuss schwer an Gebäude und Kuppel beschädigt. Kurz darauf hatte die Rote Armee das Gebäude besetzt und gleichzeitig wurde mit dem Wiederaufbau durch Kienle begonnen. Auf Befehl aus Moskau wurden aber eine ganze Reihe von technischen Geräten demontiert. Kienle jedoch konnte aber auch unter den Sowjets noch weiterarbeiten, und entschied sich nicht zur Flucht in die Zonen der westlichen Alliierten.[45]

5.6 Heidelberg – DDR – Heidelberg: Kienles Zeit 1945 bis 1962

Die Zeit Hans Kienles nach dem Kriegsende in der DDR bis 1950 lässt sich durch viele Quellen aus dem Landeshauptarchiv Potsdam, aus dem Universitätsarchiv Heidelberg, aus dem Archiv der Technischen-Universität Braunschweig, aus dem Archiv der Berlin-Brandenburgischen Akademie der Wissenschaften Berlin, aus dem Archiv des Deutschen Museums, dem Archiv der Max-Planck-Gesellschaft zur Förderung der Wissenschaften e.V., und auch aus den Beständen des Bundesarchivs, gut rekonstruieren, benötigt aber aufgrund der großen Masse an Überlieferungsgut viel an Forschungszeit.

Es ist historische Tatsache, dass die Ruprecht-Karls-Universität Heidelberg schon im Herbst 1945 auf Kienle zukam, den dortigen Lehrstuhl für Astronomie und die Sternwarte am Königstuhl im amerikanischen Sektor zu übernehmen, und man ihn aus Potsdam abwerben wollte. Hier standen unzweifelhaft die

43 Bundesarchiv Berlin, Bestand R 4901/13268, Reichsministerium für Erziehung, Wissenschaft und Forschung, Karteikarte zu Hans Kienle, Sig. K142.

44 Universitätsarchiv Heidelberg, Personalakte Hans Kienle, Signatur PA 4481; Bericht von Hans Kienle über seine Zeit während des Dritten Reiches, mss. Dok., 2.S.

45 Buthmann: Versagtes Vertrauen. Wissenschaftler der DDR im Visier der Staatssicherheit, 2020, S. 619 f.

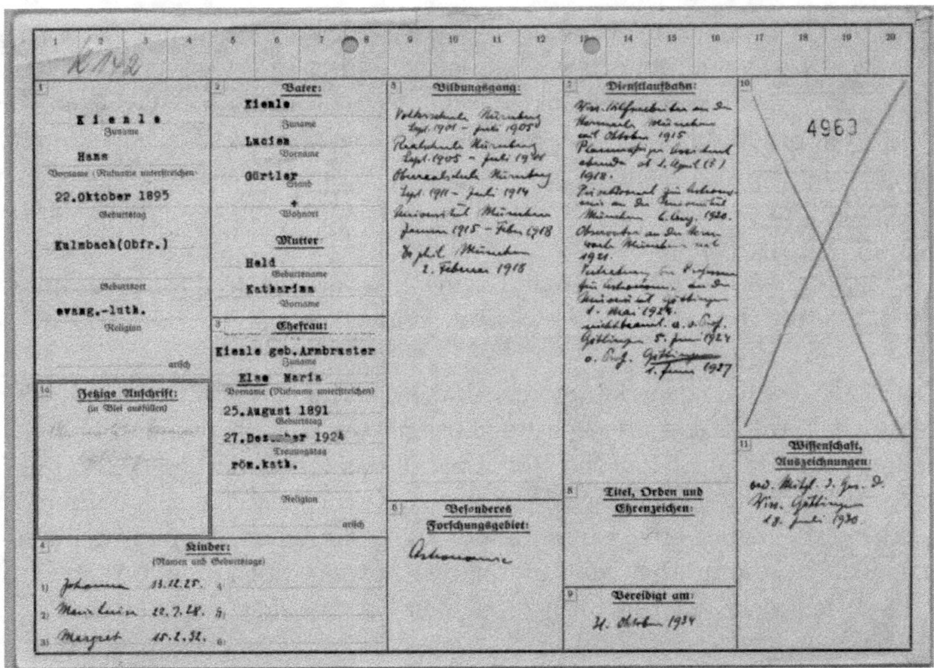

K142

1 Kienle
Zuname
Hans
Vorname (Rufname unterstreichen)
22.Oktober 1895
Geburtstag
Kulmbach(Obfr.)
Geburtsort
evang.-luth.
Religion
arisch

2 Vater:
Kienle
Zuname
Lucien
Vorname
Gürtler
Stand
†
Wohnort
Mutter:
Held
Geburtsname
Katharina
Vorname

3 Ehefrau:
Kienle geb.Armbruster
Zuname
Else Maria
Vorname (Rufname unterstreichen)
25.August 1891
Geburtstag
27.Dezember 1924
Trauungstag
röm.kath.
Religion
arisch

1a Jetzige Anschrift:
(in Blei ausfüllen)

4 Kinder:
(Namen und Geburtstage)
1) Johanna 13.12.25.
2) Marie Luise 22.7.28.
3) Margret 15.2.32.

5 Bildungsgang:
Volksschule Nürnberg Sept. 1901 – Juli 1905
Realschule Nürnberg Sept. 1905 – Juli 1911
Oberrealschule Nürnberg Sept. 1911 – Juli 1914
Universität München Januar 1915 – Febr. 1918
Dr. phil. München 2. Februar 1918

6 Besonderes Forschungsgebiet:
Astronomie

7 Dienstlaufbahn:
Wiss. Hilfsarbeiter an der Sternwarte München seit Oktober 1915
[illegible] ab 1. April (?) 1918.
Privatdozent für Astronomie an der Universität München 6. Aug. 1920.
Observator an der Sternwarte München seit 1921.
Berufung als Professor für Astronomie an die Universität Göttingen 1. Mai 1924.
nichtbeamt. a. o. Prof. Göttingen 5. Juni 1924
o. Prof. Göttingen 1. Juni 1927

8 Titel, Orden und Ehrenzeichen:

9 Vereidigt am:
31. Oktober 1934

10
4963

11 Wissenschaft, Auszeichnungen:
ord. Mitgl. d. Ges. d. Wiss. Göttingen 18. Juli 1930

Abbildung 5.4:
Karteikarte zu Hans Kienle, geführt vom Reichsministerium für Erziehung, Wissenschaft und Forschung

(Bundesarchiv Berlin, Bestand R 4901/13268, Reichsministerium für Erziehung, Wissenschaft und Forschung, Karteikarte zu Hans Kienle, Sig. K142)

Amerikaner im Hintergrund, die einen Wissenschaftler seines Formats unter ihren Einfluss bringen wollten, wie sie es u. a. auch mit Wernher von Braun erfolgreich taten. Naturwissenschaftler, insbesondere Physiker und Astronomen, sollten ihnen im aufkommenden Konflikt mit der Sowjetunion zur Verfügung stehen. Ein zügiges Entnazifizierungsverfahren durch die amerikanische Militärregierung stufte ihn deshalb schnell als unbelastet ein, so dass der Rektor der Universität Heidelberg am 12. Februar 1946 nach Potsdam an Kienle schrieb und ihn bat *„ich ersuche Sie, sobald als möglich das Amt als Ordinarius zu übernehmen.“*[46]

Doch Kienle griff nicht nach dieser ersten Gelegenheit, sich wie viele andere der ihm bekannten Akademiker aus dem Sowjetsektor zu flüchten. Die genauen

46 Universitätsarchiv Heidelberg, Personalakte Hans Kienle, Signatur PA 4481; Rektor der Universität Heidelberg an Hans Kienle am 12. Februar 1946, mss. Dok., 2 S.

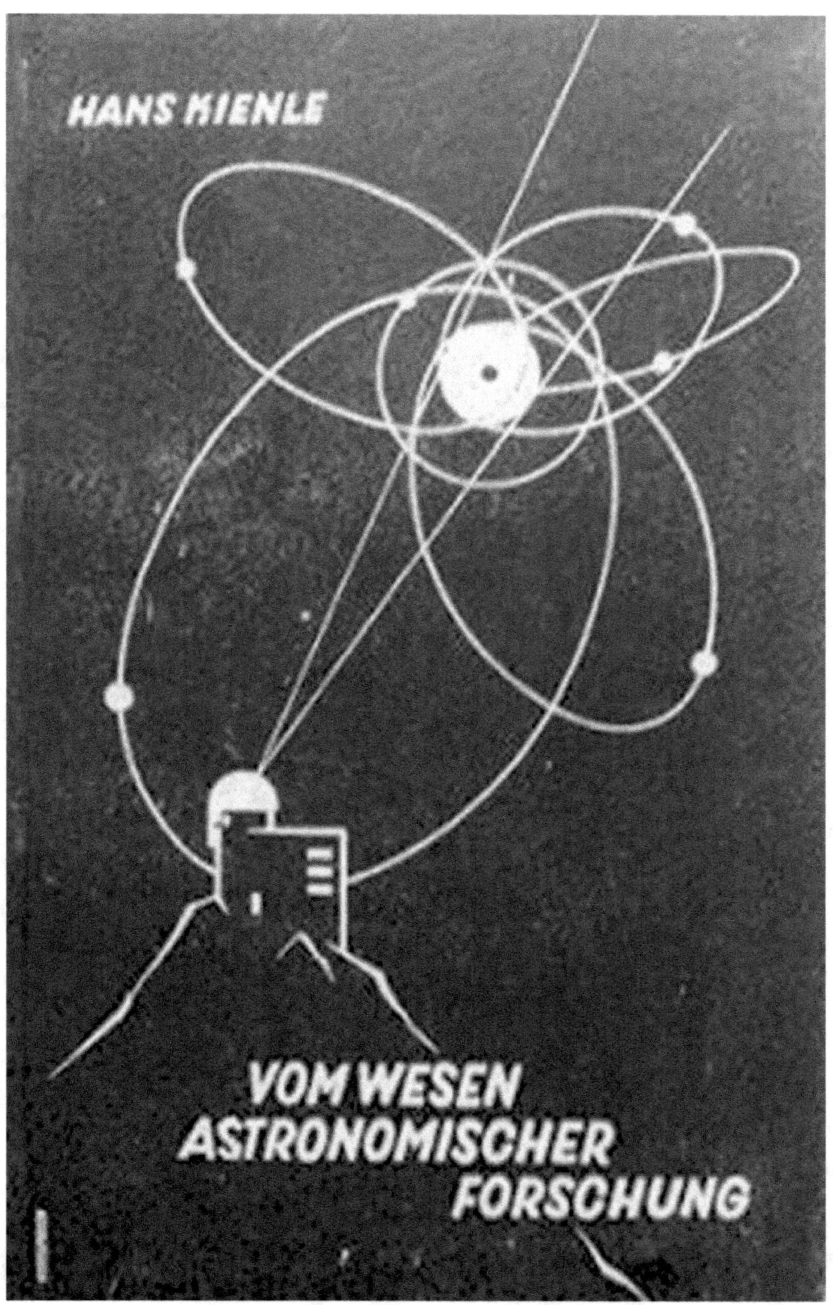

Abbildung 5.5:
Ein bedeutendes Buch von Hans Kienle:
Vom Wesen Astronomischer Forschung (1948)

Gründe dafür sind noch hinreichender zu erforschen, warum er diese Gelegenheit Heidelberg nicht sofort wahrnahm. Ein Grund war aber, dass er in seiner Position als Leiter des Potsdamer Observatoriums verbleiben konnte, und den Sowjets jeder Akademiker, und vor allem seines Formats, wichtig war. Schon bald wurde ihm klar, dass ein Verbleib in Potsdam seiner Karriere offensichtlich nicht schaden könnte, hatte er in Potsdam noch viel Unerledigtes abzuarbeiten. Sein Plan, den er schon mehrere Jahre lang verfolgte, aber wegen des Krieges nicht vollenden konnte, war der Entwurf eines großen Spiegelteleskops bei den Carl Zeiss-Werken in Jena.[47] Auch bekam er schnell Anerkennung im sowjetisch kontrollierten Wissenschaftssystem der späteren DDR. So wurde er am 1. August 1946 in die wiederbegründete „Akademie der Wissenschaften" aufgenommen, die an diesem Tag ihre Tätigkeit nach dem Krieg wieder aufnahm. Er hielt sogar den festlichen Vortrag, und zwar über die Maßstäbe des Kosmos.[48] Auch redigierte er ab 1947 die *„Astronomischen Nachrichten"* von Potsdam aus, und zwar bis zu der Ausgabe 1951.[49]

Kienle hatte mit der akademischen Lehre in Berlin und dem Wiederaufbau des Observatoriums Potsdam und dem Wunsch des Baus eines Spiegelteleskops viel zu tun. Kienle betreute dieses Vorhaben später auch nach seiner Berufung nach Heidelberg 1950 weiter, auch über die Grenzen im geteilten Deutschland hinweg. Er war ein Beispiel eines Wissenschaftlers, der scheinbar zwischen den beiden Systemen Ost und West „grenzenlos" hin und her wandeln konnte, und von diesen auch in seiner Position anerkannt wurde.[50]

Der ihm gut bekannte Genetiker Hans Stubbe (1902–1989), der ebenfalls in der DDR verblieb, schrieb um 1950 an Kienle, niemand aus der BRD möge ihnen beiden irgendeinen Vorwurf machen „dass ich noch hier bin".[51] Dennoch schien Kienle zu dieser Zeit mit dem Verbleib in der DDR gehadert zu haben. Ein wichtiger Grund war nach seinen Angaben gewesen, dass kaum ein Forscher aus den westlichen Demokratien aufgrund der kommunistischen Ideologie bereit wäre, an sein Institut nach Potsdam zu kommen und dort mit ihm zu arbeiten, denn viele seiner Wunschkandidaten hatten bereits die DDR verlassen, solange es ihnen noch möglich war.[52]

Dann kam Anfang 1950 wieder ein Angebot aus Heidelberg, dem er sich dieses Mal nicht versagen konnte, und der Übernahme der Professur an der

47 Frankfurter Allgemeine Zeitung vom 27. Oktober 1965, Artikel Hans Kienle 70 Jahre.
48 Wirth: Weltanschauliche und wissenschaftstheoretische Aspekte im Werk Hans Kienles, 2000, S. 151–168.
49 Vgl. Ausgaben der Astronomischen Nachrichten 1947–1951.
50 Lemke: Im Himmel über Heidelberg, 2019, S. 30.
51 Buthmann: Versagtes Vertrauen, 2020, S. 299.
52 Buthmann: Versagtes Vertrauen, 2020, S. 640 f.

alten Heidelberger Universität zustimmte. Im September 1950 ernannte ihn der Präsident des Landesbezirks Baden für Kultus und Unterricht *„auf Lebenszeit zum planmäßigen ordentlichen Professor für Astronomie an der Universität Heidelberg und zum Direktor der Landessternwarte auf dem Königsstuhl.“*[53]

Dennoch blieb Kienle der Astronomie in der DDR und der Sowjetunion weiter verbunden, und so auch dem Ausbau der astronomischen Forschung in der DDR.[54] So nahm er teil an der Einweihung des 2 m-Spiegelteleskops des Karl-Schwarzschild-Observatorium Tautenburg bei Jena am 19. Oktober 1960, das für die astronomische und astrophysikalische Forschung in den nächsten Jahrzehnten auch über die DDR hinaus große Bedeutung bekam. Damit ging auch sein langer Traum in Erfüllung.[55] Ein Jahr später im Oktober 1961 nahm Kienle an einem Empfang des Präsidenten der Deutschen Akademie der Wissenschaften in Berlin teil. Dort sollen sich teilnehmende DDR Wissenschaftler angeregt mit ihm unterhalten haben, und einige sollen der Meinung gewesen sein, Kienle sei gerade erst in die BRD „übergesiedelt“. Auch wenn dem nicht so war, zeigte es den Respekt, den man vor Kienle als einen führenden deutschen und internationalen Astronomen auch auf kommunistischer Seite zollte.[56]

Dass aber Hans Kienle mit den Bedingungen in Heidelberg nicht immer zufrieden war, legen verschiedene Quellen offen. So zweifelte im November 1957 ernsthaft an den Forschungsbedingungen der Landessternwarte Königstuhl, und dachte darüber nach, die Universität zu verlassen. Es ging um verwaltungs- und verfahrenstechnische Schwierigkeiten seitens des Kultusministeriums, wodurch sich Kienle in seiner Handlungsfähigkeit gehemmt sah.[57] Nur der persönliche Einsatz des Rektors beim Kultusministerium konnte ein weiteres Verbleiben Kienles bis 1962 bewirken.[58] Wie wichtig er für die Universität war, zeigt auch, dass er seit 1953 viele Jahre lang der Präsident der *Heidelberger Akademie der Wissenschaften* gewesen war.[59] Hans Kienle erhielt auch 1960 die Ehrung zum „Ritter des Ordens Pour le Mérite“ und leitete die Landessternwarte am Königstuhl von 1950 bis 1962. Er entwickelte mit seinen Mitarbeitern – als wichtiger Beitrag zur Astrophysik – die Spektralphotometrie von Sternen und der

53 Universitätsarchiv Heidelberg, Personalakte Hans Kienle, Signatur PA 4482; Personalakte, Präsident des Landesbezirks Baden für Kultus und Unterricht an den Rektor der Universität Heidelberg am 16. Oktober 1950, mss. Dok., 1 S.

54 Lemke: Im Himmel über Heidelberg, 2019, S. 30.

55 Archiv der Leopoldina, Matrikelmappe Kienle, Flyer der Einweihung des 2 m-Spiegelteleskops des Karl-Schwarzschild-Observatorium Tautenburg am 19. Oktober 1960.

56 Buthmann: Versagtes Vertrauen, 2020, S. 176.

57 Universitätsarchiv Heidelberg, Personalakte Hans Kienle, Signatur PA 4482; Personalakte, Kienle an den Kultusminister Banden Württemberg am 14.11.1957, mss. Dok., 1 S.

58 Universitätsarchiv Heidelberg, Personalakte Hans Kienle, Signatur PA 4482; Personalakte, Rektor an den Kultusminister Baden Württemberg am 14.11.1957, mss. Dok., 1 S.

59 Frankfurter Allgemeine Zeitung vom 27. Oktober 1965, Artikel Hans Kienle 70 Jahre.

HANS KIENLE

dankt

für die Glückwünsche, die ihm zum 65. Geburtstag in so überreicher Zahl zugegangen sind.

Das Meer von Blumen, in das er sich bei der Rückkehr von Tautenburg versenkt sah, die mannigfachen Gaben materieller und geistiger Art, begleitet von Briefen, in denen Anerkennung der Leistungen eines vierzigjährigen Wirkens im Dienste der Wissenschaft sich verbindet mit Bekundungen menschlicher Verbundenheit und Wünschen für eine noch lange Erhaltung von Temperament und Schaffenskraft – all das läßt die Überschreitung dieser Wegmarke in beglückendem Licht erscheinen.

Die Übergabe des 2 m-Spiegelteleskops in Tautenburg leitet den neuen Lebensabschnitt ein, in dem der homo ludens sich ganz dem hingeben darf, wozu es ihn drängt.

Abbildung 5.6:
Flyer zur Gratulation zu Hans Kienles 65. Geburtstag

(Archiv der Leopoldina, Matrikelmappe Kienle,
Flyer zur Gratulation zu Hans Kienles 65. Geburtstag)

Sonne in hoher Präzision. Während seines Direktorates wurde auf dem Gelände der Sternwarte das *Happel-Laboratorium für Strahlungsmessungen* Arbeits- und Gästezimmer errichtet.

Als im Jahre 1962 Hans Kienle im Alter von 67 Jahren emeritiert wurde, übernahm der aus Göttingen berufene 33-jährige Hans Elsässer (1929–2003) seine Nachfolge.[60] Kienle hatte das Glück, ein volles berufliches Leben der Astronomie zu widmen, von Beginn seines Studiums 1915 bis zu seinem Ruhestand 1962, und damit fast ein halbes Jahrhundert lang.

60 Lemke: Im Himmel über Heidelberg, 2019, S. 428 f.

Abbildung 5.7:
Flyer zur Gratulation zu Hans Kienles 65. Geburtstag mit Bildern seiner Teilnahme an der Einweihung des Zeiss Spiegelteleskops in Tautenburg am 19. Oktober 1961

(Archiv der Leopoldina, Matrikelmappe Kienle, Flyer zur Gratulation zu Hans Kienles 65. Geburtstag)

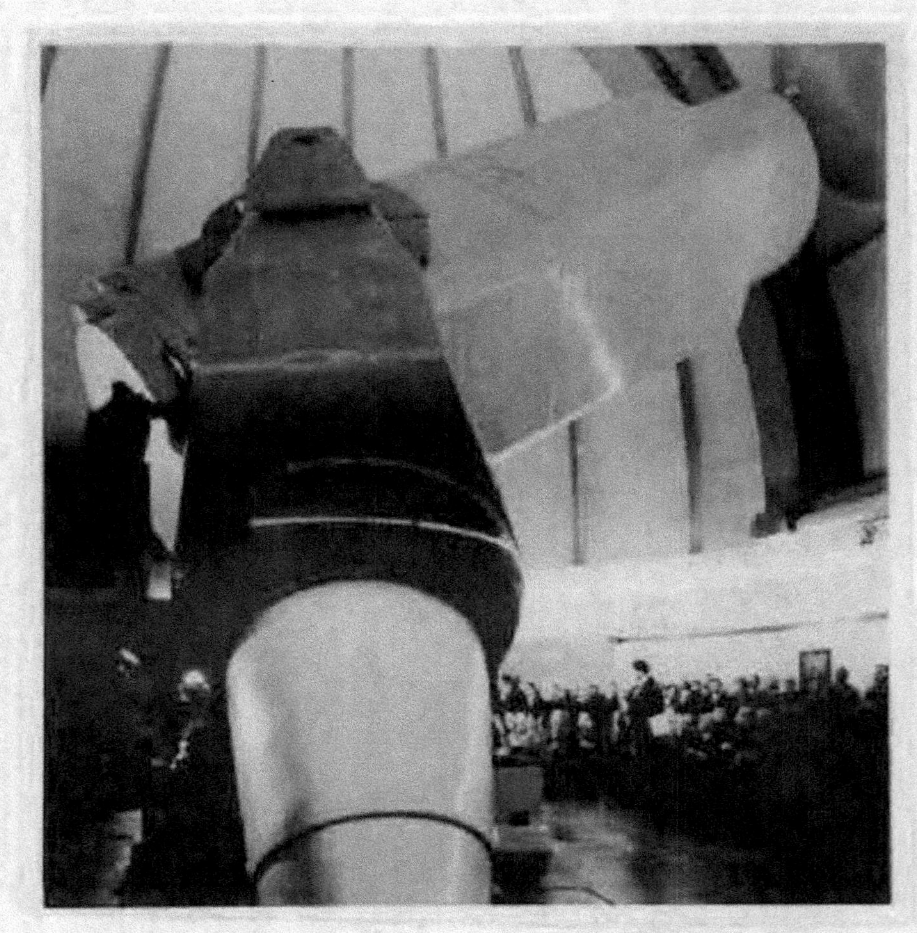

2 m-Spiegelteleskop
des Karl-Schwarzschild-Observatoriums Tautenburg
der Deutschen Akademie der Wissenschaften zu Berlin
bei der feierlichen Übergabe am 19. 10. 1960

Abbildung 5.8:
Flyer zur Gratulation zu Hans Kienles 65. Geburtstag mit Bildern seiner Teilnahme an der Einweihung des Zeiss Spiegelteleskops in Tautenburg am 19. Oktober 1961

(Archiv der Leopoldina, Matrikelmappe Kienle,
Flyer zur Gratulation zu Hans Kienles 65. Geburtstag)

5.7 Ruhestand und Tod

Als er 1967 in den Ruhestand ging, bedeutete das für Kienle aber keineswegs, sich der astronomischen Forschung zu entziehen. Denn einige andere Projekte wurden ihm angetragen. Von 1962 bis 1964 richtete er im Auftrag der UNESCO einen Lehr- und Forschungsbetrieb in einer ägyptischen Sternwarte in Kotamya ein. Anschließend wurde er als Gastprofessor sogar für viele Jahre nach Izmir in die Türkei eingeladen, wo er ein Observatorium einrichtete und Vorlesungen an der Universität hielt. Auch die zahlreichen beruflichen Verbindungen pflegte er weiter.[61]

Am 15. Februar 1975 starb Hans Kienle nach einem langen Leben für die Astronomie, und nach den Abschiedsworten seiner Frau Elsa *„als mein lieber Mann, unser guter Vater und Opa“*. Ihm war Zeit seines Lebens beides wichtig gewesen: sein Beruf und seine Familie.[62] Werner Heisenberg (1901–1976) gedachte ihm mit sehr ausführlichen „Gedenkworte für Hans Kienle“ in der Zeitschrift des Ordens.[63]

Eine private, aber vor allem die berufliche Biographie über Kienle sollte zu allen Stationen seines Lebens und zur Astronomie und Astrophysik des 20. Jahrhunderts zahlreiche neue Erkenntnisse bringen, auch zur Universitätsgeschichte des Faches selbst. Es bleibt abzuwarten, wie lange eine Forschung in Anspruch nehmen wird. Sein 50. Todestag im Jahr 2025 wäre dafür ein geeignetes Gedenkjahr.

5.8 Literatur

BARTH, HANS: *Hermann Oberth – der wirkliche Vater der Weltraumfahrt.* Düsseldorf 2008.

BUTHMANN, REINHARD: *Versagtes Vertrauen. Wissenschaftler der DDR im Visier der Staatssicherheit.* Göttingen 2020, S. 619 f.

ELSÄSSER, HANS (Einführung): *Lexikon der Astronomie. Die große Enzyklopädie der Weltraumforschung in zwei Bänden.* Erster Band A bis Mirzam. Freiburg: Herder 1995.

GEYKEN, FRAUKE: *Zum Wohle Aller. Geschichte der Georg-August-Universität Göttingen von ihrer Gründung 1737 bis 2019.* Göttingen: Steidl Verlag 2019.

HECKMANN, OTTO: Nachruf auf Hans Kienle. In: *Mitteilungen der Astronomischen Gesellschaft* 38 (1976), S. 9–11.

61 Frankfurter Allgemeine Zeitung vom 27. Oktober 1965, Art.: Hans Kienle 70 Jahre.
62 Archiv der Leopoldina, Matrikelmappe Kienle, Todesanzeige Hans Kienle 1975.
63 Heisenberg: Gedenkworte für Hans Kienle 1975, S. 128–134.

Kienle

Abbildung 5.9:
Bild von Hans Kienle mit dem Orden Pour le Mérite

(Orden Pour le Mérite für Wissenschaften und Künste. Die Mitglieder des Ordens, Dritter Band, Die Verstorbenen der Jahre 1953–1992. Gerlingen 1994, S. 97)

HEISENBERG, WERNER: Gedenkworte für Hans Kienle. In: *Orden Pour Le Merite für Wissenschaften und Künste, Reden und Gedenkworte, Band 12.* Heidelberg 1975, S. 128–134.

HERRMANN, DIETER B.: Friedrich Simon Archenhold und seine Treptower Sternwarte. In: *Vorträge und Schriften der Archenhold-Sternwarte Berlin*, Nr. 65 (1986).

HOFFMANN, HORST: *Sigmund Jähn – Rückblick ins All: die Biografie des ersten deutschen Kosmonauten.* Berlin 2008.

International Who is Who 1967–1968. London: Europa publications (31th edition) 1968.

KIENLE, HANS: Jüngerer Astronom [...]. In: *Astronomische Nachrichten*, Band **227** (1926), S. 271.

KIENLE, HANS: *Vom Wesen Astronomischer Forschung – Aufsätze und Vorträge.* Berlin: Aufbau-Verlag 1948.

KILIAN, ULRICH & ROLF SAUERMOST (Red.): *Brockhaus der Astronomie. Planeten-Sterne-Galaxien.* Mannheim: F. A. Brockhaus 2005.

KUMMER, HANS JOCHEN: Hans Kienle. Ein Lebensbild zu seinem 100. Geburtstag. In: *Sterne und Weltraum* **35** (1996), Heft 4, S. 266–269.

LEMKE, DIETRICH: *Im Himmel über Heidelberg. 50 Jahre Max-Planck-Institut für Astronomie in Heidelberg (1969–2019).* Berlin, Heidelberg 2019.

LÜBCKE, ANGELIKA: Alice Archenhold (1874–1943): Hausfrau, Mitarbeiterin in der Astronomie (mitarbeitende Ehefrau). In: *Frauenmosaik. Frauenbiographien aus dem Berliner Stadtbezirk Treptow-Köpenick.* Hg. vom Bezirksamt Treptow-Köpenick von Berlin. Berlin 2002, S. 13–20.

MESTEL, LEON: Martin Schwarzschild. In: *The Royal Society. Biographical Memoirs of Fellows of the Royal Society* (1999), S. 469–484.

PEDDE, FRIEDHELM: Verfolgung und Exil. Verfemte deutsche Astronomen im Dritten Reich. Teil 3. Martin Schwarzschild, Friedrich und Günter Archenhold. In: *... dem Himmel nahe ... der Erde verbunden. Mitgliederzeitschrift der Wilhelm-Foerster-Sternwarte, Berlin*, Ausgabe 10, März-April-Mai 2021, S. 8–11.

REIF, SIEGLINDE: *Hunger, Wohnungsnot, Arbeitslosigkeit, ‚Demobilmachung'. Folgen des Ersten Weltkrieges in München.* München 1995.

REINSCH, KLAUS & AXEL D. WITTMANN (Hg.): *Karl Schwarzschild (1873–1916): ein Pionier und Wegbereiter der Astrophysik.* Göttingen 2017.

STRACH, ERIC: Obituary Gunter Archenhold, 1904–1999. In: *Journal of the British Astronomical Association* 109 (1999), 4, S. 226.

WIRTH, GÜNTER: Weltanschauliche und wissenschaftstheoretische Aspekte im Werk Hans Kienles. In: DICK, WOLFGANG R. & KLAUS FRITZE (Hg.): *300 Jahre Astronomie in Berlin und Potsdam. Eine Sammlung von Aufsätzen aus Anlaß des Gründungsjubiläums der Berliner Sternwarte.* Frankfurt am Main: Harri Deutsch (Acta Historica Astronomiae; Band 8) 2000, S. 151–168.

Abbildung 6.1:
Einsteinturm mit Kriegsschaden

(Staatliche Museen Preußischer Kulturbesitz, Kunstbibliothek Berlin, Mendelssohn-Nachlaß)

Vertreibung und Emigration von Astrophysikern im Nationalsozialismus

Stefan L. Wolff (München)

Abstract: Expulsion and Emigration of Astrophysicists in National Socialism

While in mathematics, sciences, medicine, and also in technology, the extent of the dismissals of scientists in the Nazi state is now quantitatively quite well recorded, the situation in astronomy is more difficult because it was not primarily organised as a university subject. Astronomers sometimes held academic titles such as 'titular professor', but without being active in teaching. With the fairly well-documented personnel changes at the universities, astronomy can therefore not be recorded, knowledge of the number of staff at the observatories would be required.

However, we have chosen a different path here. Using the personnel lists of the emigrant aid organisations, we identify a total of 16 dismissed scientists who we can assign to astronomy in Germany, Austria, and Prague, but only ten of them worked at a scientific observatory. The others are two university lecturers who came to astronomy as physicists at a technical university (Robert Emden (1862–1940)) or were in the process of devoting themselves to astronomy (Rudolf Minkowski (1895–1976)), a director of a public observatory (Günter Archenhold (1904–1999)), an astronomer who worked as a journalist without a permanent position (Arthur Beer (1900–1980)), a director of a university training observatory (Hermann von Socher (1892–1980)), and a young scientist who had to leave Germany immediately after completing his doctorate (Martin Schwarzschild (1912–1997)). Apart from a special situation in Kiel, where the two astronomers and university lecturers working there lost their jobs, the dismissals, which were formally carried out mostly, but not only, on the basis of the Professional Civil Servants Act (*Berufsbeamtengesetz*, BBG), did not otherwise lead to such visible disruptions as in the other sciences.

Of the ten dismissed astronomers who worked at observatories, five were because of their own background (Erwin Freundlich (1885–1964), Adolph Henry Rosenthal

(1906–1962), Richard Prager (1883–1945), Hans Rosenberg (1879–1940), Wolfgang Gleißberg (1903–1986)), three because of the origin of their partners (Carl Wilhelm Wirtz (1876–1939), Hermann Brück (1905–2000), Rupert Wildt (1905–1976), and two for political reasons (Alexander Wilkens (1881–1968), Kasimir Graff (1878–1950)) affected by the sanctions. Only two of the five dismissed (Freundlich and Rosenberg) held a managerial position.

While university departments increasingly opened up to scientists from Jewish families in the late 19th century, this was obviously far less the case in the observatories with a manageable number of paid positions. Of the total of 16 astronomers affected, 14 emigrated, with Istanbul being an option for continuing their scientific work, in addition to the United Kingdom and the USA. In the lecture, some of these individual emigration fates with their adaptation problems were described.

Zusammenfassung

Während in der Mathematik, den Naturwissenschaften, der Medizin und auch in der Technik das Ausmaß der Entlassungen von WissenschaftlerInnen im NS-Staat inzwischen quantitativ recht gut erfasst wird, stellt sich die Situation in der Astronomie als schwieriger dar, weil sie nicht hauptsächlich als Universitätsfach aufgestellt war. So führten Astronomen mitunter zwar akademische Titel wie den „Titularprofessor", ohne aber in der Lehre tätig zu sein. Mit den recht gut dokumentierten personellen Veränderungen an den Universitäten kann man die Astronomie demnach nicht erfassen, stattdessen wären Kenntnisse über den Personalbestand der Sternwarten erforderlich.

Wir wählen hier aber einen anderen Weg. Über die Personallisten der Emigrantenhilfsorganisationen identifizieren wir insgesamt 16 entlassene Wissenschaftler, die wir der Astronomie in Deutschland, Österreich und Prag zuordnen können, davon gehörten aber nur zehn zu denen, die an einer wissenschaftlichen Sternwarte arbeiteten. Die anderen sind zwei Hochschullehrer, die als Physiker an der Technischen Hochschule/Universität zur Astronomie kamen (Robert Emden (1862–1940)) bzw. dabei waren, sich der Astronomie zuzuwenden (Rudolf Minkowski (1895–1976)), ein Leiter einer Volkssternwarte (Günter Archenhold (1904–1999)), ein Astronom, der ohne feste Anstellung publizistisch tätig war (Arthur Beer (1900–1980)), ein Leiter einer universitären Übungssternwarte (Hermann von Socher (1892–1980)) sowie ein Nachwuchswissenschaftler, der gleich nach der Promotion Deutschland verlassen musste (Martin Schwarzschild (1912–1997)). Abgesehen von einer Sondersituation in Kiel, wo die beiden dort tätigen Astronomen und Hochschullehrer ihre Stellen verloren, führten die Entlassungen, die formal meist, aber nicht nur auf der Grundlage des *Berufsbeamtengesetzes* (BBG) durchgeführt wurden, ansonsten zu keinen so sichtbaren Brüchen wie in den anderen Naturwissenschaften.

Von den zehn entlassenen Astronomen, die an Sternwarten arbeiteten, waren fünf aufgrund ihrer eigenen Herkunft (Erwin Freundlich (1885–1964), Adolph Henry Rosenthal (1906–1962), Richard Prager (1883–1945), Hans Rosenberg (1879–1940), Wolfgang Gleißberg (1903–1986)), drei aufgrund der Herkunft ihrer Partnerinnen (Carl Wilhelm Wirtz (1876–1939), Hermann Brück (1905–2000), Rupert Wildt (1905–1976) sowie zwei aus politischen Gründen (Alexander Wilkens (1881–1968), Kasimir Graff (1878–1950)) von den Maßnahmen betroffen. Nur zwei der fünf aufgrund der Herkunft Entlassenen (Freundlich und Rosenberg) übten dabei eine leitende Funktion aus.

Während die Universitätsfächer, sich im späten 19. Jahrhundert auch zunehmend für Wissenschaftler aus jüdischen Familien öffneten, war das in den Sternwarten mit einer überschaubaren Zahl von bezahlten Stellen offenbar weitaus weniger der Fall. Von den insgesamt 16 betroffenen Astronomen sind 14 emigriert, wobei es neben dem Vereinigten Königreich und den USA noch Istanbul als Option für eine Fortsetzung der wissenschaftlichen Arbeit gab. In dem Vortrag wurden einige dieser individuellen Emigrationsschicksale mit ihren Adaptionsproblemen geschildert.

6.1 Literatur

Wolff, Stefan L.: Vertreibung und Emigration in der Physik. In: Physik in unserer Zeit 24 (1993), S. 267–273.

Wolff, Stefan L.: Die Ausgrenzung und Vertreibung von Physikern im Nationalsozialismus – welche Rolle spielte die DPG? In: Hoffmann, Dieter & Mark Walker (ed.): *Physiker zwischen Autonomie und Anpassung.* Berlin: Wiley 2006, S. 91–138.

Wolff, Stefan L.: Marginalization and Expulsion of Physicists under National Socialism: What was the German Physical Society's Role? In: Hoffmann, Dieter & Mark Walker: The German Physical Society and the Third Reich. Physicists between Autonomy and Accommodation. Cambridge Mass.: Cambridge University Press 2012, p. 50–95.

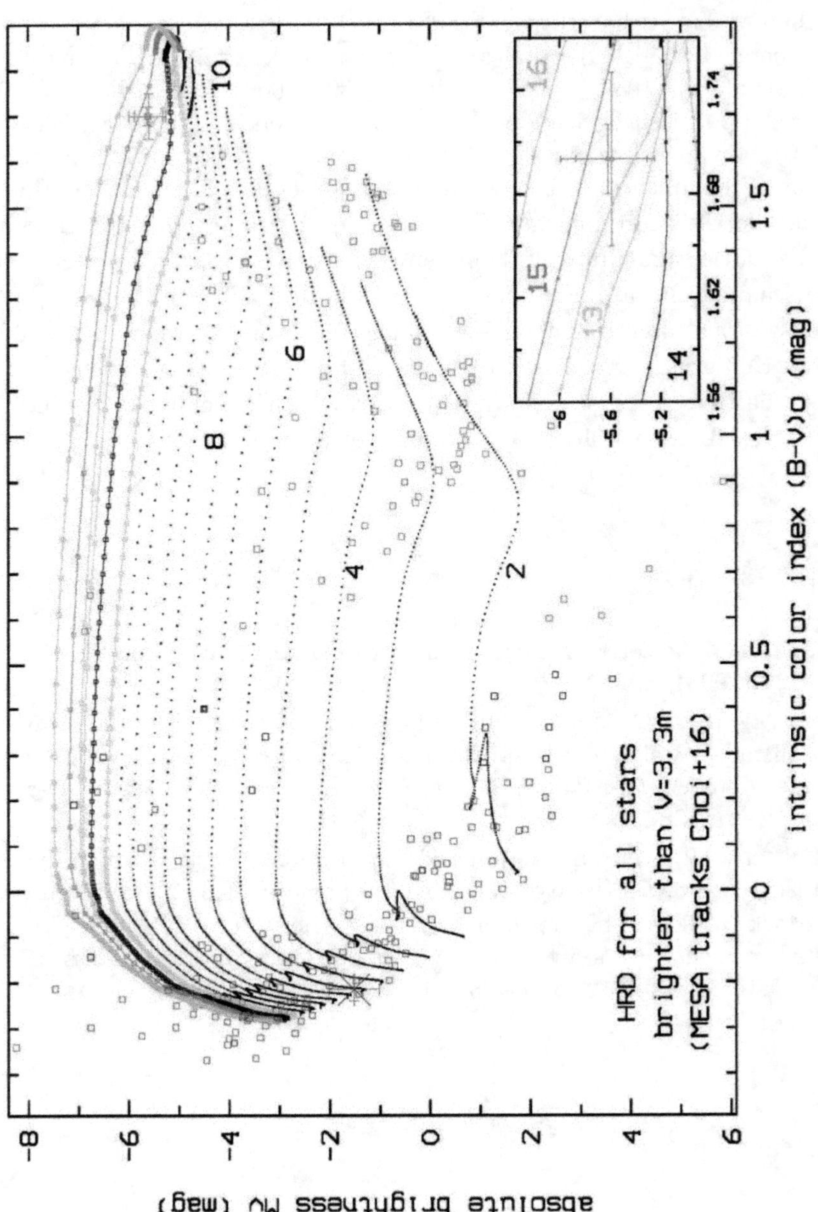

Abbildung 7.1:
Farb-Helligkeits-Diagramm für alle Sterne bis V=3.3 mag mit MESA-MIST Entwicklungswegen für verschiedene Massen. Beteigeuze und Antares sind im Ausschnitt rechts unten vergrössert nochmal dargestellt. Beteigeuze hat die Hertzsprung-Lücke gerade verlassen.

(Neuhäuser et al. (2022))

Die rasche Evolution von Beteigeuze durch die Hertzsprung-Lücke vom gelben zum roten Überriesen in historischer Zeit

Ralph Neuhäuser (Jena) & Dagmar L. Neuhäuser (Meran, Italien)

Abstract: Rapid evolution of Betelgeuse through the Hertzsprung gap from yellow to red in historical times

After core hydrogen burning, massive stars evolve rapidly from blue-white main sequence dwarfs to red giants or super-giants - by expanding and cooling. For stars with some 8 to 18 solar masses, the evolution is as rapid as ca. 10,000 to 100,000 years (Kippenhahn 1965), the so-called Hertzsprung gap: Due to rapid evolution, only few stars reside in this area of the Hertzsprung-Russell diagram (HRD) or color-magnitude diagram (CMD), hence the gap. It may therefore be possible to notice a color change from blue-white over yellow to red by crossing the Hertzsprung gap in historical time.

We have considered the 236 brightest stars down to 3.3 mag, for which the human naked eye can detect color: We have placed them into the CMD in order to study those where rapid color evolution in historical time is possible. Among those super-giants are Betelgeuse at the end of the Hertzsprung gap and Antares, which has already passed the gap; both are bright, deep red, nearby super-giants, where the naked eye can clearly see red color.

In parallel, we have compiled pre-telescopic color reports on stars from many cultures including Europe, Arabia, and East Asia. The oldest records on stars colors include Antares and Mirach as red in from the 2nd millennium BC (China and Mul.apin).

The star Betelgeuse (α Ori) was compared by Hyginus (Rome, BC/AD turn) in his work on astronomy with the color of Saturn: *"Saturn ... is in color as the star on the right shoulder of Orion"*, which is of course Betelgeuse, because the constellation figures look at us. Saturn as planet is constant in color and has a color index of $B - V = 1.09 \pm 0.16$ mag. Hence, Betelgeuse had that color index two millennia ago. Also, the Chinese astronomer Sima Qian (ca. 100 BC, Han dynasty) wrote: *"Venus white like Lang (Sirius), red like Xin (Antares), yellow like the left shoulder of Shen (Betelgeuse), blue like to right shoulder of Shen (Bellatrix) ... "*; in China, the upper left shoulder on sky is seen as Betelgeuse. The color sequence is Bellatrix blue, Sirius white, Betelgeuse yellow, and then Mirach and Antares in red. Apparently, Sima meant that Sirius and Bellatrix had a lower color index compared to Betelgeuse, while that of Antares and Mirach was larger. A value between the blue-white Sirius and Bellatrix and the red Antares, Mirach and Mars (also given as red) is then 0.95 ± 0.35 mag. Today, Betelgeuse has a color index of $B - V = 1.78 \pm 0.05$ mag, significantly larger than two millennia ago. Both sources from antiquity are fully consistent with each other.

The color change of Betelgeuse to a red super-giant within the last two millennia is not only just consistent with theoretical evolutionary tracks (e. g. MESA MIST tracks), but is also confirms and constrains them even further; from these tracks and the color evolution constraint, we can determine the mass of Betelgeuse to be around 14 solar masses. The remaining life-time is then ca. 1.5 Myr until the supernova. (Previously, mass and age of Betelgeuse was more uncertain given its uncertain distance and, hence, luminosity, because its apparent diameter is larger than its parallax.)

Zusammenfassung

Nach dem Wasserstoffbrennen im Kern entwickeln sich massereiche Sterne rasch von blau-weissen Zwergsternen zu Roten Riesen oder Überriesen – durch Expansion und Kühlung. Für Sterne mit ca. 8 bis 18 Sonnenmassen erwartet man eine Durchquerung der Hertzsprung-Lücke innerhalb von nur ca. 10,000 Jahren (z. B. Kippenhahn 1965). Da sich Sterne in diesem Bereich des Hertzsprung-Russell- bzw. Farb-Helligkeits-Diagramms schnell entwickeln, gibt es nur wenig Sterne bei diesen Parametern, daher die Lücke. Es ist möglich, dass eine Farbänderung von hellen Sternen bei der Durchquerung der Hertzsprung-Lücke in historischer Zeit festzustellen ist.

Wir haben alle 236 Sterne, bei denen man mit blossem Auge im Prinzip die Farbe erkennen kann, d. h. bis ca. 3,3 mag, in das Hertzsprung-Russell- bzw. Farb-Helligkeits-Diagramm eingetragen, um diejenigen weiter zu untersuchen, die z. Z. (oder bis vor kurzem) in der Hertzsprung-Lücke liegen (bzw. lagen). Dies gilt u. a. für Beteigeuze und Antares, die heute als helle, sehr rote Überriesen bekannt sind.

Ebenso haben wir historische Berichte über Farben von Sternen zusammengetragen, um festzustellen, ob bei einigen Objekten eine Farbe genannt wird, die heute nicht mehr zutreffend ist. Die Sterne Antares (α Sco) und Mirach (β And) wurden bereits im 2. Jahrtausend v. Chr. als rot bezeichnet (China bzw. in Mul.apin).

Der Stern Beteigeuze (α Ori) jedoch wurde von Hyginus (Rom, um die Zeitenwende) als in der Farbe wie Saturn beschrieben: *„Saturn ... ist in Farbe wie der Stern in der rechten Schulter des Orion.“* Saturn – als Planet konstant in Farbe – hat einen Farbindex $B - V = 1,09 \pm 0,16$ mag; und damit kann Beteigeuzes damalige Farbe bestimmt werden. Ferner hat der chinesische Astronom Sima Qian (um 100 v. Chr., Han-Dynastie) geschrieben: *„Venus weiss wie Lang (Sirius), rot wie Xin (Antares), gelb wie Shen linke Schulter (Beteigeuze), blau wie Shen rechte Schulter (Bellatrix) ... “*. In der Tat sind Sirius weiss, Antares rot, Bellatrix blau. Somit liegt Beteigeuze zwischen diesen Sternen im Farbindex; zusammen mit weiteren Informationem ergibt sich $B - V = 0,95 \pm 0,35$ mag. Heute hat Beteigeuze jedoch einen Farbindex von $B - V = 1,78 \pm 0,05$ mag, also höchst signifikant anders als vor ca. 2000 Jahren. Die Angaben dieser zwei unabhängigen Quellen von Hyginus und Sima Qian sind miteinander und auch mit weiteren Quellen der Antike konsistent.

Die Farbänderung von Beteigeuze zu einem roten Überriesen in den letzten Jahrtausenden ist nicht nur voll konsistent mit theoretischen Entwicklungswegen, sondern sie bestätigt diese und schränkt sie weiter ein; man erwartet bei den MESA-Modellen für einen Stern mit 14 Sonnenmassen, der jetzt $B - V = 1,78 \pm 0,05$ mag hat, dass er in den letzten Jahrtausenden sich von einem gelben zu einem roten Überriesen entwickelt hat. Damit läßt die Masse von Beteigeuze durch die historisch belegte Farbänderung präzise bestimmen – und somit auch seine Restlebenszeit bis zur Supernova in erst ca. 1,5 Millionen Jahren. (Vorher waren Masse und Alter stark unsicher, da der Durchmesser von Beteigeuze als sehr naher Überriesen grösser als seine Parallaxe ist.)

7.1 Literatur

Neuhäuser Dagmar L. & Ralph Neuhäuser: Les couleurs de Betelgeuse – L'intérêt des observations prételescopiques pour l'astrophysique moderne. In: *L'Astronomie* **170** (2023), p. 38–45.

Neuhäuser, Dagmar L.: Beteigeuze – Vom Gelben zum Roten Überriesen – Vorteleskopische Beobachtungen als Erkenntnisschlüssel. In: *Sterne und Weltraum* **62** (2023), Heft 1, p. 23–25.

Neuhäuser, Ralph; Torres, G.; Mugrauer, M.; Neuhäuser, Dagmar L.; Chapman, J.; Luge, D. & M. Cosci: Colour evolution of Betelgeuse and Antares over two millennia, derived from historical records, as a new constraint on mass and age. In: *Monthly Notices of the Royal Astronomical Society* (MNRAS) **516** (2022), p. 693–719, doi.org/10.1093/mnras/stac1969.

Abbildung 8.1:
Oben: James-Webb-Space-Telescope (JWST)
Unten: Start des James-Webb-Space-Telescopes
mit der europäischen Ariane 5 Rakete am 25. Dez. 2021

(ESA, C. Carreau, ESA)

Erlebte Geschichte – Das James-Webb-Space-Telescope – Von der Idee zur Mission

Dietrich Lemke (Heidelberg)

Abstract:

With a 6.5-meter primary mirror, it is the largest space observatory ever launched. Two of the four complex scientific instruments on board were developed in Europe. The 25-year development history of this project was characterized by scientific curiosity, technological innovations and broad international cooperation; but also accompanied by crises, cost increases and delays. The James-Webb-Space-Telescope (JWST) is now proving its unique capabilities by providing unprecedented views of the universe, such as galaxies in the early cosmos, the earliest phases of star formation, and young planetary systems around nearby stars.

Zusammenfassung

Mit einem 6,5-m-Hauptspiegel ist es das größte Weltraum-Observatorium das je gestartet wurde. Zwei der vier komplexen wissenschaftlichen Instrumente an Bord wurden in Europa entwickelt. Die 25 Jahre währende Entwicklungsgeschichte dieses Vorhabens war geprägt durch wissenschaftliche Neugier, technologische Innovationen und breite internationale Zusammenarbeit; aber auch begleitet von Krisen, Kostensteigerungen und Verzögerungen. Das James-Webb-Space-Telescope (JWST) beweist jetzt seine einzigartige Leistungsfähigkeit durch nie gesehene Blicke in das Universum, so auf Galaxien im frühen Kosmos, auf die frühesten Phasen der Sternentstehung und auf junge Planetensysteme bei nahen Sternen.

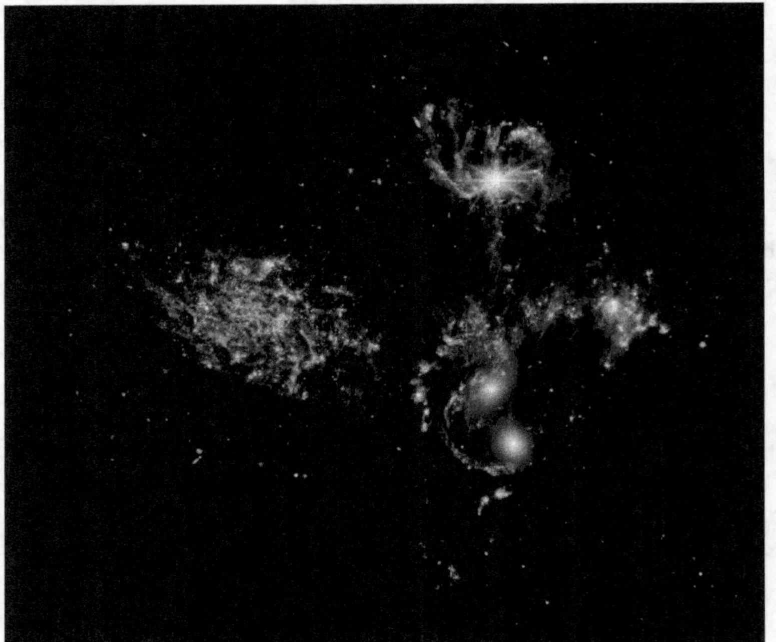

Abbildung 8.2:
Above: Carina Nebula: Behind the curtain of dust and gas in these *Cosmic Cliffs* are previously hidden baby stars, now uncovered by Webb.
Below: Stephan's Quintet: These colliding galaxies – 5 galaxies, 4 of which interact – are pulling and stretching each other in a gravitational dance.

(Credits: NASA, ESA, CSA, and STScI, `nasa.gov/webbfirstimages/`)

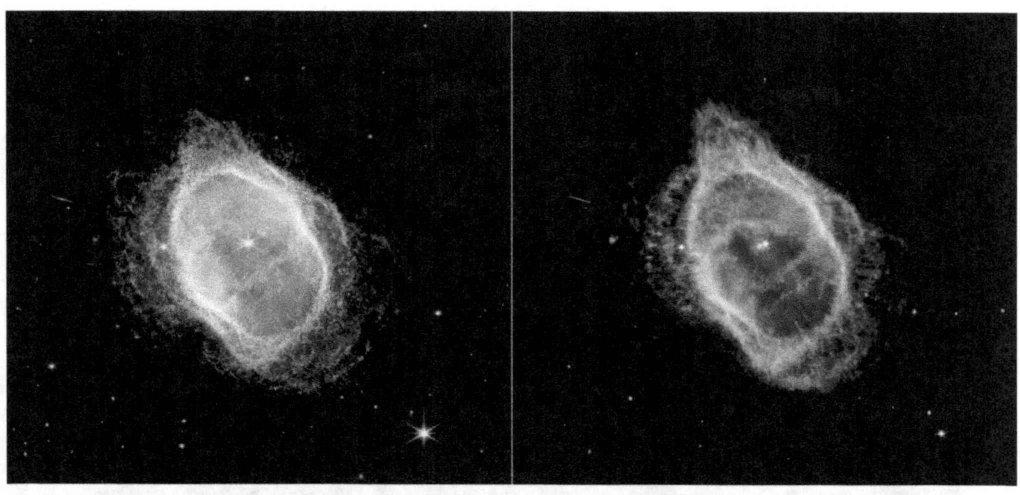

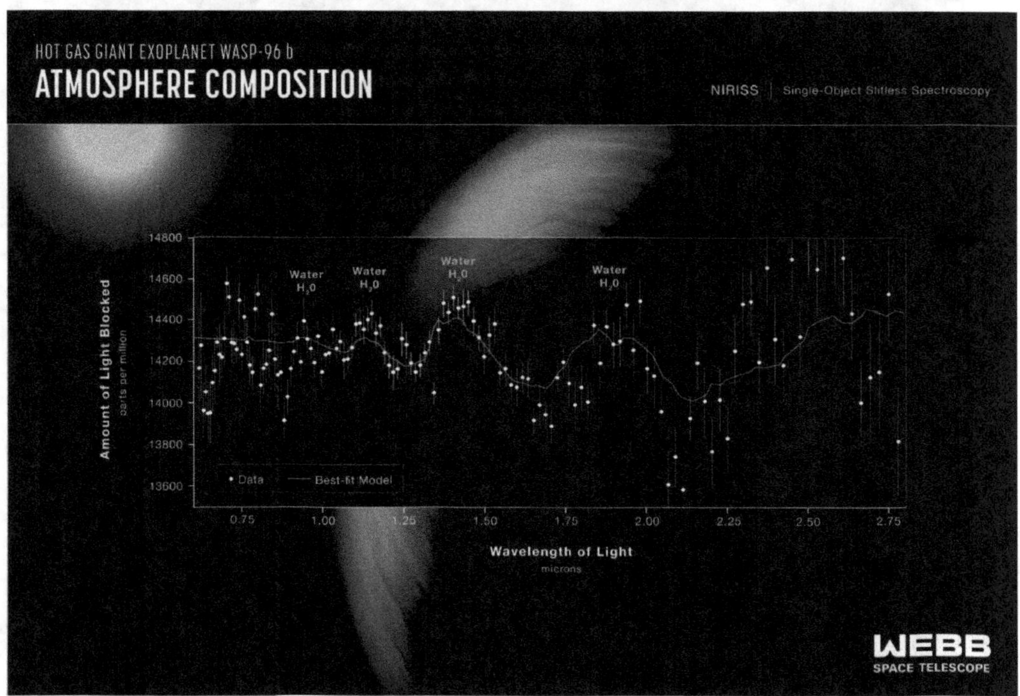

Abbildung 8.3:
Above: Dying Star's Final 'Performance': A dying star expelling gas and dust, in orbit with a younger star that is helping to change the shape of this nebula's intricate rings by creating turbulence.
Below: Clouds are in the forecast for exoplanet WASP-96 b: The James Webb Space Telescope spotted the unambiguous signature of water, indications of haze & evidence for clouds – the most detailed exoplanet spectrum.

(Credits: NASA, ESA, CSA, and STScI, `nasa.gov/webbfirstimages/`)

Abbildung 9.1:
Wilhelm-Förster-Sternwarte (WFS) Berlin,
31,4 cm Refraktor (Brennweite 5 m), Carl Bamberg, Berlin-Friedenau (1889)

Foto: Gudrun Wolfschmidt

Links – Astronomie, Museen in Berlin

Gudrun Wolfschmidt (Hamburg)

9.1 Allgemeine Links zur Astronomie und Astronomiegeschichte

- Arbeitskreis Astronomiegeschichte in der Astronomischen Gesellschaft: (http://www.astronomische-gesellschaft.de/de/arbeitskreise/Astronomiegeschichte/Veranstaltungen).

- Nuncius Hamburgensis, hg. von Gudrun Wolfschmidt: (https://www.fhsev.de/Wolfschmidt/GNT/research/nuncius.php).

- Acta Historica Astronomiae, Publikationsreihe des Arbeitskreises Astronomiegeschichte, hg. von Wolfgang R. Dick & Jürgen Hamel. Leipzig: AVA – Akademische Verlagsanstalt. (http://www.univerlag-leipzig.de/catalog/category/158-Acta_Historica_Astronomiae).

- Annual meeting 2023 of the Astronomische Gesellschaft (AG) (https://ag2023.astronomische-gesellschaft.de/program/index.php): *Cosmic Evolution of Matter on all Scales.* Technical University Berlin, September 11–15, 2023.

9.2 Links zur Astronomie und ihrer Geschichte in Berlin

- Zentrum für Astronomie und Astrophysik der Technischen Universität Berlin
  Hardenbergstr. 36, 10623 Berlin
  (https://www-astro.physik.tu-berlin.de/)
- Geschichte: Astronomie im geteilten Berlin,
  (https://www-astro.physik.tu-berlin.de/node/2)
  Mitglieder des Zentrums bzw. Instituts
  (https://www-astro.physik.tu-berlin.de/node/182)
- Archenhold-Sternwarte (*1896),
  Alt-Treptow 1, 12435 Berlin
  (https://www.planetarium.berlin/archenhold-sternwarte/geschichte)
- Zeiss-Großplanetarium Berlin,
  Prenzlauer Allee 80, 10405 Berlin
  (https://www.planetarium.berlin/zeiss-grossplanetarium)
  S-Bahn: S41, S42, S8, S85 – Prenzlauer Allee
  Tram: Tram M2 – Fröbelstraße, Tram M10 – Prenzlauer Allee/Danziger Straße
  Bus: 156 – S Prenzlauer Allee (Grellstraße)
- Wilhelm-Foerster-Sternwarte Berlin - Planetarium am Insulaner
  Munsterdamm 90 – 12169 Berlin
  (https://www.planetarium.berlin/planetarium-am-insulaner-wilhelm-foerster-sternwarte)
  S-Bahn: S2, S25, S26 – Priesterweg
  Bus: 187 – Planetarium (z.Z. geschlossen), M76, 170, 246 – Insulaner
- Leibniz-Institut für Astrophysik Potsdam
  An der Sternwarte 16, 14482 Potsdam-Babelsberg
  (http://www.aip.de/)
- Potsdam-Telegrafenberg, Historical Site:
  (https://www.aip.de/en/institute/locations/the-potsdam-telegrafenberg-historical-site)
  History: (https://www.aip.de/en/institute/history)
  Large Refractor: (https://www.aip.de/en/press/gf-et)
  Einsteinturm Potsdam:
  (https://www.aip.de/de/institute/locations/einstein-tower/).

Abbildung 9.2:
Deutsches Technikmuseum Berlin – „Spektrum“

Foto: Gudrun Wolfschmidt

9.3 Museen in Berlin

(Auswahl: besonders Naturwissenschaft, Technik, Kulturgeschichte)

- Deutsches Technikmuseum Berlin – „Spektrum“
  Trebbiner Str. 9, 10963 Berlin-Kreuzberg
  (`https://sdtb.de/technikmuseum/startseite/`)
  Universitätsmuseen – TU Berlin
  (`http://www.universitaetssammlungen.de/search/uni/Technische+Universit%C3%A4t+Berlin`)
  Universitätsmuseen – Humboldt Universität Berlin
  (`http://www.universitaetssammlungen.de/search/uni/Humboldt-Universit%C3%A4t+zu+Berlin`)

Abbildung 10.1:
Astrophysikalisches Observatorium Potsdam (1874–1879),
heute Michelson-Haus des Potsdam Instituts für Klimafolgenforschung (PIK)

(Foto: Gudrun Wolfschmidt, 2012)

Tagung des Arbeitskreises Astronomiegeschichte in Berlin 2023

Colloquium of the Working Group History of Astronomy in the Astronomical Society (AKAG), Berlin 2023

10.0.1 SOC – Scientific Organizing Committee

- Prof. Dr. Gudrun Wolfschmidt – Chair
  (University of Hamburg)
- Dr. Stefan L. Wolff (München)
- Dr. cand. Karsten Markus-Schnabel, M.Sc. (Lübeck)

10.0.2 LOC – Local Organizing Committee

- Prof. Dr. Gudrun Wolfschmidt
  (Hamburg)
- Dr. Stefan L. Wolff (München)
- Dr. cand. Karsten Markus-Schnabel, M.Sc. (Lübeck)

10.1 Sonntag, 10. September 2023, 15 Uhr, Berlin AKAG Tagung – Exkursion: Archenhold-Sternwarte Berlin

- Exkursion zur Archenhold-Sternwarte (*1896),
  Alt-Treptow 1, 12435 Berlin.

  Nachmittag: 15 Uhr – Beginn der Führung von
  PD Dr. Felix Lühning
  (Grosser Refraktor und Ausstellung).

  Anfahrt mit S-Bahn:
  S8, S9 – Plänterwald (12 min Fußweg)
  S41, S42 – Treptower Park (20 min Fußweg)
  Anfahrt mit Bus:
  165, 166, 265, N65 – Alt-Treptow

- 20 Uhr – Angebot zu einem gemeinsamen Abendessen –
  DAS LEMKE – Brauerei & Deutsche Küche,
  Dircksenstr., S-Bahnbogen 143, 10178 Berlin-Mitte

Abbildung 10.2:
68 cm-Refraktor (21 m Brennweite) in der Archenhold-Sternwarte Treptow, Riesenfernrohr – das *längste* voll bewegliche Linsenfernrohr der Welt, Steinheil, München, C. Hoppe, Berlin

Foto: Gudrun Wolfschmidt

10.2 Berlin, Montag, 11. September 2023

Technische Universität – TU Berlin

Adresse: Straße des 17. Juni 135, 10623 Berlin,
Konferenzraum: Hauptgebäude (H 3025)
(https://www.tu-berlin.de/menue/home/)

09:00 – 9:20 Uhr – Registration / Anmeldung

09:20 – 10:30 Uhr – Session 1: Einführung: Von der klassischen Astronomie zur theoretischen Astrophysik

Chair: **Stefan L. Wolff (München)**

- 09:20 Uhr–09:30 Uhr – Grußworte – Welcome
- 09:30 Uhr–10:00 Uhr – Prof. Dr. Gudrun Wolfschmidt (Hamburg): *Einführung zum Thema: Von der klassischen Astronomie zur theoretischen Astrophysik: 150 Jahre – Karl Schwarzschild (1873–1916) und Ejnar Hertzsprung (1873–1967) – Pioniere der theoretischen Astrophysik*
- 10:00–10:30 Uhr – Markus Bautsch (Berlin): *Die von Johann Jakob Balmer gefundenen Zahlenverhältnisse bei den Spektrallinien des Wasserstoffs*

10:30–11:00 Uhr – Kaffeepause – Coffee Break

10:30–11:00 Uhr – Session 2: Astrophysik in der 1. Hälfte des 20. Jahrhunderts

Chair: **Karsten Markus-Schnabel (Lübeck)**

- 11:00–11:30 Uhr – Ralph Neuhäuser (Jena) & Dagmar L. Neuhäuser (Meran, Italien): *Die rasche Evolution von Beteigeuze durch die Hertzsprung-Lücke vom gelben zum roten Überriesen in historischer Zeit*
- 11:30–12:00 Uhr – Maik Schmerbauch (Berlin): *Hans Kienle (1895–1975) – Wissenschaftler zwischen Astrophysik und Politik im 20. Jahrhundert*

12:00–14:00 Uhr – Mittagessen – Lunch Break
Diverse Lokale an der Straße des 17. Juni 135, 10623 Berlin

14:00–16:30 Uhr – Session 3: Astrophysik ab Mitte des 20. Jahrhunderts

Chair: **Gudrun Wolfschmidt (Hamburg)**

- 14:00–15:00 Uhr – Dietrich Lemke (Heidelberg)
  Erlebte Geschichte – Das James-Webb-Space-Telescope – Von der Idee zur Mission
- 15:00–15:30 Uhr – Stefan L. Wolff (München):
  Vertreibung und Emigration von Astrophysikern im Nationalsozialismus

15:30–16:00 Uhr – Kaffeepause – Coffee Break

- 16:00–16:30 Uhr – Xian Wu (Dresden):
  Heinrich Kaysers „Handbuch der Spectroscopie“ (1900–1934) und dessen Bedeutung für die Astrophysik seit 1900
- – Uhr – Rita Meyer-Spasche (Garching, München):
  Astronomie und Astrophysik nach den zwei Weltkriegen: Wiederbelebung und Strukturwandel des internationalen wissenschaftlichen Austausches
  leider verhindert.

16:30–18:00 Uhr – Mitgliederversammlung des Arbeitskreises Astronomiegeschichte (AKAG)

Donnerstag, 14. September 2023 – TU Berlin, H 0104

- 09:00–9:30 Uhr – Dr. Adriaan Raap (Königsbronn):
  Karl Schwarzschild and Ejnar Hertzsprung in Potsdam (1910 bis 1916)

Abbildung 10.3:
Grosser Refraktor (80 cm ph.-gr., 50 cm vis.), Steinheil, München, Repsold & Söhne, Hamburg (1899), Astrophysikalisches Observatorium Potsdam

(Foto: Gudrun Wolfschmidt, 2012)

List of Participants – Astrophysik seit 1900 – AKAG Berlin 2023

1. Bautsch, Markus, Dr. Dipl.-Phys. (Berlin): (drbautsch@aol.com)
2. Bolze, Günter Paul (Wien, Österreich): (bolzegp@yahoo.de)
3. Dick, Wolfgang R., Dr. (Potsdam) (wdick@astrohist.org)
4. Hänel, Andreas, Dr. (Osnabrück) (ahaenel@uos.de) – verhindert
5. Hoffmann, Susanne M., Dr. (Jena) (akademeia@exopla.net)
6. Kitmeridis, Panagiotis, Dr. (Frankfurt am Main) (kitmeridis@t-online.de) – verhindert
7. Lemke, Dietrich, Prof. Dr. (MPIA, Heidelberg) (lemke@mpia-hd.mpg.de)
8. Markus-Schnabel, Karsten, Master of Science in Astronomy, Dr.cand. (Lübeck, GNT Universität Hamburg), (karsten.markus@gmail.com)
9. Mattila, Kalevi, Prof. Dr. (Helsinki, Finnland) (mattila@cc.helsinki.fi) – verhindert
10. Meyer-Spasche, Rita, PD Dr. (MPI für Plasmaphysik (IPP), Garching) (rita.meyer-spasche@ipp.mpg.de) – verhindert
11. Neuhäuser, Dagmar, M.A. (Jena)
12. Neuhäuser, Ralph, Prof. Dr. (Astrophysikalisches Institut und Universitäts-Sternwarte Jena) (ralph.neuhaeuser@uni-jena.de)
13. Raap, Ignatius Adriaan, Dr.rer.nat. (Königsbronn) (dr.araap@gmail.com)
14. Reinsch, Klaus, Dr. (Universität Göttingen) (reinsch@astro.physik.uni-goettingen.de)
15. Schmerbauch, Maik, Dr. Dr. (Berlin): (schmeichi@web.de)
16. Umland, Regina (Mannheim) (Umland@t-online.de) – verhindert

17. Wolff, Stefan L., Dr. (München) (s.wolff@deutsches-museum.de)
18. Wolfschmidt, Gudrun, Prof. Dr.
(GNT, Hamburger Sternwarte, Uni Hamburg)
(gudrun.wolfschmidt@uni-hamburg.de)
19. Wu, Xian, Dr. (Dresden)
(wuxn03@hotmail.com)

Abbildung 11.1:
Zeiss-Großplanetarium Berlin (1987), 30 m-Kuppel

Foto: Gudrun Wolfschmidt, Zeiss-Großplanetarium, Prenzlauer Allee 80, 10405 Berlin

Autoren

Dr. Dipl.-Phys. Markus Bautsch (Berlin))

1964 in Berlin geboren. Die Schwerpunktgebiete beim Studium der Physik und der Geodäsie an der Technischen Universität Berlin waren neben der Teilchenoptik auch die Astrophysik, die numerische Mathematik und die Informatik. Beruflich ist er als Wissenschaftler für die Konzeption und Durchführung von internationalen und vergleichenden Untersuchungen in den Bereichen Digitales und Technik verantwortlich. Er unterrichtet seit Jahrzehnten an verschiedenen Hochschulen in den Fächern Physik, Elektrotechnik, Optik, Informatik oder Mechanik.
Er ist Mitglied im Verein der Wilhelm-Foerster-Sternwarte in Berlin und betätigt sich auch in der Archäoastronomie. In seiner Freizeit engagiert er sich ferner als Sänger und Chorleiter sowie als Wikipedia-Autor.

Berlin
`https://de.wikipedia.org/wiki/Benutzer:Bautsc` E-Mail: `DrBautsch@AOL.com`

Prof. Dr. Dietrich Lemke (Heidelberg)

Dietrich Lemke leitete bis zu seiner Emeritierung am Max-Planck-Institut für Astronomie in Heidelberg verschiedene Weltraumprojekte.

Königstuhl 17, 69117 Heidelberg, Germany
E-Mail: `lemke@mpia-hd.mpg.de`

Dagmar Neuhäuser (Jena)

Astrophysikalisches Institut und Universitäts-Sternwarte Jena
E-Mail: `ralph.neuhaeuser@uni-jena.de`

Prof. Dr. Ralph Neuhäuser (Jena)

Astrophysikalisches Institut und Universitäts-Sternwarte Jena
E-Mail: `ralph.neuhaeuser@uni-jena.de`

Dr. Ignatius Adriaan Raap (Königsbronn)

Königsbronn
E-Mail: dr.araap@gmail.com

Dr. Dr. Maik Schmerbauch (Berlin)

Dr. phil., Dr. theol. Maik Schmerbauch (Jahrgang 1979), hat Geschichte, Philosophie, Theologie, Informationswissenschaften und Archivistik studiert.
Er leitet das Archiv der Katholischen Militärseelsorge in Berlin. Forschungsschwerpunkte sind Kirchengeschichte, Militärgeschichte sowie Archiv- und Bibliotheksgeschichte im 19. und 20. Jahrhundert. Als langjähriger Hobbyastronom forscht er auch zur Geschichte der Astronomie im 20. Jahrhundert.

Berlin
E-Mail: schmeichi@web.de

Dr. Stefan L. Wolff (München)

Physikhistoriker am Forschungsinstitut des Deutschen Museums in München

Deutsches Museum München
E-Mail: s.wolff@deutsches-museum.de

Prof. Dr. Gudrun Wolfschmidt (Hamburg)

Dissertation *Analyse enger Doppelsternsysteme*, FAU – Friedrich-Alexander-Universität Erlangen-Nürnberg (Remeis-Sternwarte Bamberg), 1. und 2. Staatsexamen (Physik und Mathematik), 1987–1997 Deutsches Museum in München: Ausstellung Astronomie (Eröffnung 1992, bis 2022) und Assistentin am Forschungsinstitut für Geschichte der Naturwissenschaft und Technik, Lehre und Habilitation *Genese der Astrophysik* (1997) an der Ludwig-Maximilians-Universität in München (LMU).
Seit 1997 Professorin für Geschichte der Naturwissenschaft und Technik an der Universität Hamburg, Center for History of Science and Technology.
Aktivitäten siehe: https://www.fhsev.de/Wolfschmidt/
Forschungsschwerpunkte:
Astronomie-, Physik-, Wissenschafts- und Technikgeschichte (Frühe Neuzeit und 19./ 20. Jahrhundert), Archäo- und Kulturastronomie, Sternwarten und wissenschaftliche Instrumente.
Buchveröffentlichungen (Auswahl):
Copernicus – Revolutionär wider Willen (1994), *Milchstraße – Nebel – Galaxien.*

Strukturen im Kosmos von Herschel bis Hubble (1995), *Popularisierung der Naturwissenschaften / Astronomie* (2000, 2002, 2017), *Sterne weisen den Weg – Geschichte der Navigation* (2009), *Cultural Heritage of Astronomical Observatories* (2009), *Colours in Culture and Science* (2011), *Harmony and Symmetry* (2020), *Astronomy in Culture – Cultures of Astronomy* (2022).
Publikationen – Gesamt-Liste: https://www.fhsev.de/Wolfschmidt/publikat.php.
Herausgeberin der Reihe *Nuncus Hamburgensis* https://www.fhsev.de/Wolfschmidt/GNT/research/nuncius.php.

AG Geschichte der Naturwissenschaft und Technik (GNT)
Hamburger Sternwarte, Universität Hamburg
Bundesstraße 55 – Geomatikum, 20146 Hamburg
https://www.fhsev.de/Wolfschmidt/GNT/home-wf.htm
E-Mail: gudrun.wolfschmidt@uni-hamburg.de.

Dr. Xian Wu (Dresden)

geboren 1980 in Nanjing, China, studierte Materialchemie an der Technischen Universität Nanjing mit dem Abschluss B.Sc. (2002). Er setzte sein Studium in Deutschland fort und erhielt den Titel M.Sc. (2007) und Dr. rer. nat. (2011) in Chemie an der Technischen Universität Braunschweig. Anschließend erfolgte die Postdoc-Phase an der RWTH Aachen und der Universität Erlangen-Nürnberg bis 2014. Von 2014 bis 2016 arbeitete er als Regional Marketing Manager bei *American Chemical Society International, Ltd.* in Jena. Seit 2016 ist er als Geschäftsführer einer Handelsunternehmens in Jena bzw. Dresden tätig. Er erhielt den Titel M. Eng. (2019) und Dr.-Ing. (2023) in Werkstofftechnik an der Technischen Universität Bergakademie Freiberg, wo er zur Zeit als wissenschaftlicher Mitarbeiter beschäftigt ist.

Außer Chemie hat er auch großes Interesse an Astronomie, Physik, Wissenschaftsgeschichte sowie Popularisierung der Naturwissenschaften. Seine Publikationsliste ist unter orcid.org/0000-0002-1282-6972 zu finden.

Dornblüthstraße 11b, 01277 Dresden
E-Mail: wuxn03@hotmail.com

Nuncius Hamburgensis –
Beiträge zur Geschichte der Naturwissenschaften, Band 1
Hans Schimank (1888-1979)
Ausgewählte Schriften
Bearbeitet von
Timo Engels und Igor Abdrakhmanov
Norderstedt: Books on Demand 2009

Nuncius Hamburgensis –
Beiträge zur Geschichte der Naturwissenschaften, Band 2
Gudrun Wolfschmidt (Hrsg.)
Hamburgs Geschichte einmal anders
Entwicklung von Naturwissenschaft, Medizin und Technik
Norderstedt: Books on Demand 2007

Gudrun Wolfschmidt (Hg.)
Astronomie in Nürnberg

Nuncius Hamburgensis –
Beiträge zur Geschichte der Naturwissenschaften, Band 4
Gudrun Wolfschmidt (Hg.)
Entwicklung der
Theoretischen Astrophysik
tredition

Nuncius Hamburgensis –
Beiträge zur Geschichte der Naturwissenschaften, Band 12
Gudrun Wolfschmidt (Hg.)
Astronomy
in new Wavelengths
Astronomie in neuen Wellenlängen

Nuncius Hamburgensis –
Beiträge zur Geschichte der Naturwissenschaften, Band 6
Gudrun Wolfschmidt (Hrsg.)
Von Hertz zum Handy
Entwicklung der Kommunikationstechnik

Nuncius Hamburgensis –
Beiträge zur Geschichte der Naturwissenschaften, Band 7
Gudrun Wolfschmidt (Hg.)
Hamburgs Geschichte
einmal anders
Entwicklung der Naturwissenschaften,
Medizin und Technik, Teil 2

Nuncius Hamburgensis –
Beiträge zur Geschichte der Naturwissenschaften, Band 5
Prähistorische Astronomie
und Ethnoastronomie
2008

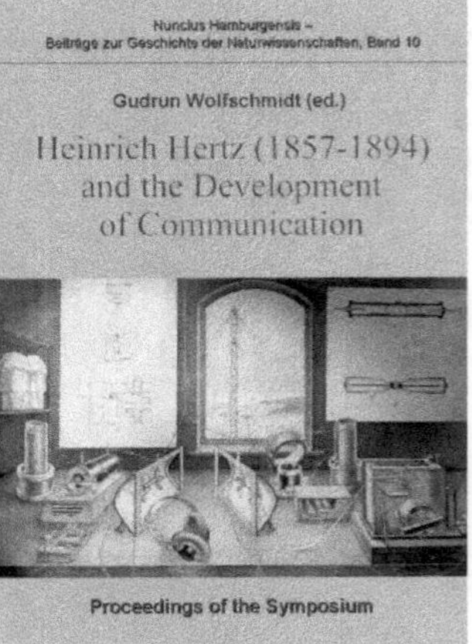
Nuncius Hamburgensis –
Beiträge zur Geschichte der Naturwissenschaften, Band 10
Gudrun Wolfschmidt (ed.)
Heinrich Hertz (1857-1894)
and the Development
of Communication
Proceedings of the Symposium

Nuncius Hamburgensis

Beiträge zur Geschichte der Naturwissenschaften

Norderstedt: Books on Demand (nur Bd. 2, 6, 7, 8, 10, 11, 14 und 15)

Hamburg: tredition Verlag  tredition® (alle anderen Bände).

Hg. von Gudrun Wolfschmidt,
AG Geschichte der Naturwissenschaft und Technik,
Hamburger Sternwarte, Universität Hamburg – ISSN 1610-6164

Diese Reihe „Nuncius Hamburgensis“ wird gefördert von der Hans Schimank-Gedächtnisstiftung. Dieser Titel wurde inspiriert von „Sidereus Nuncius“ und von „Wandsbeker Bote“.

`https://www.fhsev.de/Wolfschmidt/GNT/research/nuncius.php`

- Band 1 (2009): *Hans Schimank (1888–1979) Ausgewählte Schriften.* Mit einem Beitrag ‚Hans Schimanks Otto von Guericke‘ von Fritz Krafft. Bearb. von Timo Engels & Igor Abdrakhmanov. Hg. G. Wolfschmidt.
- Band 2 (2007): Wolfschmidt, Gudrun (Hg.): *Hamburgs Geschichte einmal anders – Entwicklung der Naturwissenschaften, Medizin und Technik – Teil 1.*
- Band 3 (2010): Wolfschmidt, Gudrun (Hg.): *Astronomie in Nürnberg.* Proceedings der Tagung vom 2.–3. April 2005 in Nürnberg anläßlich des 500. Todestages von Bernhard Walther (1430–1504) und des 300. Todestages von Georg Christoph Eimmart (1638–1705).
- Band 4 (2011): Wolfschmidt, Gudrun (Hg.): *Entwicklung der Theoretischen Astrophysik.* Proceedings des Kolloquiums des Arbeitskreises Astronomiegeschichte in der Astronomischen Gesellschaft in Köln am 26. September 2005.
- Band 5 (2026): Wolfschmidt, Gudrun (Hg.): *400 Jahre Physik in Hamburg vom Akademischen Gymnasium über das Staatsinstitut zur Universität.*
- Band 6 (2007): Wolfschmidt, Gudrun (Hg.): *Von Hertz zum Handy – Entwicklung der Kommunikation.* Begleitbuch zur Ausstellung zum 150. Geburtstag von Heinrich Hertz (1857–1894).
- Band 7 (2009): Wolfschmidt, Gudrun (Hg.): *Hamburgs Geschichte einmal anders – Entwicklung der Naturwissenschaften, Medizin und Technik, Teil 2.*

Gudrun Wolfschmidt (Hrsg.)
Astronomisches Mäzenatentum in Europa

Nuncius Hamburgensis –
Beiträge zur Geschichte der Naturwissenschaften, Band 14
Gudrun Wolfschmidt (Hrsg.)
Navigare necesse est
Geschichte der Navigation

Nuncius Hamburgensis –
Beiträge zur Geschichte der Naturwissenschaften, Band 15
Gudrun Wolfschmidt (Hrsg.)
Sterne weisen den Weg
Geschichte der Navigation

Gudrun Wolfschmidt (Hg.)
der fränkische Galilei,
und die Entwicklung
des astronomischen Weltbildes
tredition

Nuncius Hamburgensis –
Beiträge zur Geschichte der Naturwissenschaften, Band 18
Gudrun Wolfschmidt (Hg.)
Farben
in Kulturgeschichte und Naturwissenschaft

Nuncius Hamburgensis –
Beiträge zur Geschichte der Naturwissenschaften, Band 19
Andre Koch Torres Assis,
Karl Heinrich Wiederkehr
and Gudrun Wolfschmidt
Weber's Planetary Model
of the Atom
Ed. by Gudrun Wolfschmidt

Gudrun Wolfschmidt (Hg.)
Hamburgs Geschichte einmal anders
Entwicklung der Naturwissenschaften,
Medizin und Technik, Teil 3

Nuncius Hamburgensis –
Beiträge zur Geschichte der Naturwissenschaften, Band 22
Gudrun Wolfschmidt (ed.)
Colours
in Culture and Science
science

Nuncius Hamburgensis –
Beiträge zur Geschichte der Naturwissenschaften, Band 28
Gudrun Wolfschmidt (Hg.)
Der Himmel über Tübingen
Barocksternwarten -
Landesvermessung -
Hochenergieastrophysik
tredition

- Band 8 (2008): Wolfschmidt, Gudrun (Hg.): *Prähistorische Astronomie und Ethnoastronomie.* Proceedings der Tagung des Arbeitskreises Astronomiegeschichte in der AG in Würzburg 2007.
- Band 9 (2024): Wolfschmidt, Gudrun (Hg.): *Genese der Astrophysik – The Rise of Astrophysics.*
- Band 10 (2008): Wolfschmidt, Gudrun (ed.): *Heinrich Hertz (1857–1894) and the Development of Communication.* Proceedings of the International Symposium in Hamburg, Oct., 8–12, 2007.
- Band 11 (2008): Wolfschmidt, Gudrun (Hg.): *Astronomisches Mäzenatentum.* Proceedings des Symposiums in der Kuffner-Sternwarte in Wien 2004.
- Band 12 (2024): Wolfschmidt, Gudrun (Hg.): *Astronomie in neuen Wellenlängen – Astronomy in New Wavelength.* Proceedings der Tagung des Arbeitskreises Astronomiegeschichte in der Astronomischen Gesellschaft in Würzburg.
- Band 13 (2025): Cura, Katrin: *Alchemie im Deutschen Museum.* Bearb. und hg. von G. Wolfschmidt.
- Band 14 (2008): Wolfschmidt, Gudrun (Hg.): *„Navigare necesse est“ – Geschichte der Navigation.* Begleitbuch zur Ausstellung 2008/09 in Hamburg und Nürnberg.
- Band 15 (2009): Wolfschmidt, Gudrun: *„Sterne weisen den Weg“ – Geschichte der Navigation.* Katalog zur Ausstellung 2008/10 in Hamburg und Nürnberg.
- Band 16 (2012): Wolfschmidt, Gudrun (Hg.): *Simon Marius, der fränkische Galilei, und die Entwicklung des astronomischen Weltbildes.*
- Band 17 (2026): Cura, Katrin: *Auf den Leim gehen – Geschichte der Klebstoffe.* Hg. von Gudrun Wolfschmidt.
- Band 18 (2011): Wolfschmidt, Gudrun (Hg.): *Farben in Kulturgeschichte und Naturwissenschaft. Begleitbuch zur Ausstellung in Hamburg 2010–2012.*
- Band 19 (2011): Andre Koch Torres Assis und Karl Heinrich Wiederkehr und Gudrun Wolfschmidt: *Weber's Planetary Model of the Atom.*
- Band 20 (2011): Wolfschmidt, Gudrun (Hg.): *Hamburgs Geschichte einmal anders – Entwicklung der Naturwissenschaften, Medizin und Technik, Teil 3.*
- Band 21 (2019): Wolfschmidt, Gudrun (Hg.): *Vom Abakus zum Computer – Geschichte der Rechentechnik, Teil 1. Begleitbuch zur Ausstellung, 2015–2018.*
- Band 22 (2011): Wolfschmidt, Gudrun (ed.): *Colours in Culture and Science. 200 Years Goethe's Colour Theory.* Proceedings of the Interdisciplinary Symposium in Hamburg, October 12–15, 2010.

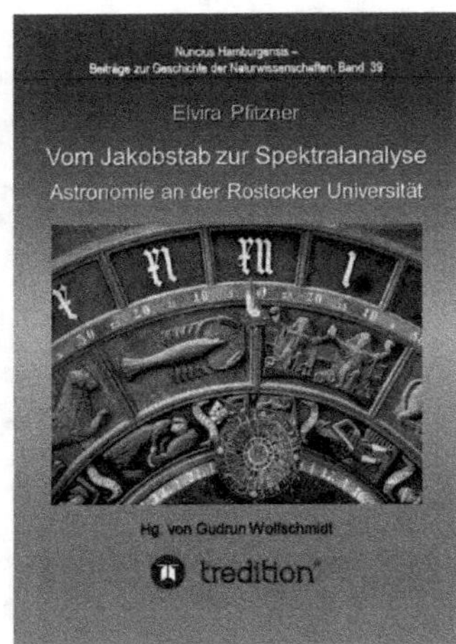
Nuncius Hamburgensis –
Beiträge zur Geschichte der Naturwissenschaften, Band 39
Elvira Pfitzner
Vom Jakobstab zur Spektralanalyse
Astronomie an der Rostocker Universität
Hg. von Gudrun Wolfschmidt
tredition

tredition
WOLFRAM ZIEGLER
AUS DEM LEBEN
EINES FRANKEN
Dr. AUGUST ZIEGLER (1885-1937)
Pflanzenzüchter in Togo
Rebenzüchter in Bayern

Nuncius Hamburgensis –
Beiträge zur Geschichte der Naturwissenschaften, Band 31
Gudrun Wolfschmidt (Hg.)
Astronomie in Franken
Von den Anfängen bis zur modernen Astrophysik.
125 Jahre Dr. Remeis-Sternwarte Bamberg (1889)
tredition

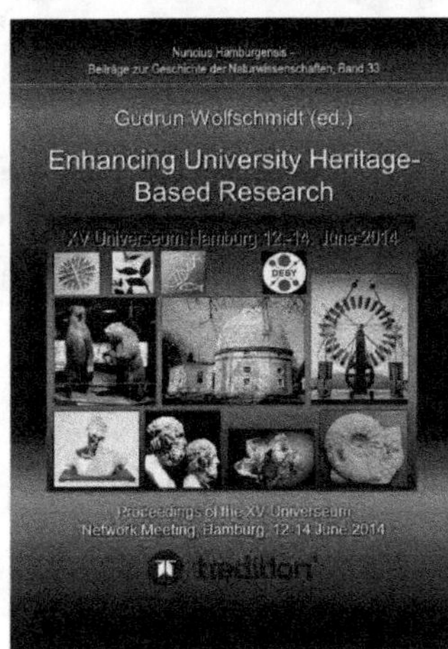
Nuncius Hamburgensis –
Beiträge zur Geschichte der Naturwissenschaften, Band 33
Gudrun Wolfschmidt (ed.)
Enhancing University Heritage-
Based Research
XV Universeum Hamburg 12.-14. June 2014
Proceedings of the XV Universeum
Network Meeting, Hamburg, 12-14 June 2014
tredition

Nuncius Hamburgensis –
Beiträge zur Geschichte der Naturwissenschaften, Band 25
Gudrun Wolfschmidt (Hg.)
400 Jahre Chemie in Hamburg
Hamburgs Geschichte einmal anders –
Entwicklung der Naturwissenschaften,
Medizin und Technik, Teil 4
tredition

Nuncius Hamburgensis –
Beiträge zur Geschichte der Naturwissenschaften, Band 43
Carlotta Martini
Zwei Frauenleben für die
Wissenschaft im 18. Jahrhundert
Eine vergleichende Fallstudie
zu Émilie du Châtelet und
Maria Gaetana Agnesi

Nuncius Hamburgensis –
Beiträge zur Geschichte der Naturwissenschaften, Band 27
Susanne M. Hoffmann (Hg.)
Lingua sine limitibus
Analysen zur Sprache der Bilder und Bildsprachen,
insbesondere zur Kommunikation von Fachinformationen
Sprachen der Populärdidaktik
mit zwei- bis vierdimensionalen Medien
an Beispielen der Astronomie
tredition

Nuncius Hamburgensis –
Beiträge zur Geschichte der Naturwissenschaften, Band 40
Jürgen Kost
Wissenschaftlicher Instrumentenbau
der Firma Merz in München (1838-1932)
tredition

Gudrun Wolfschmidt (Hg.)
Festschrift – Proceedings of the
Christoph J. Scriba Memorial Meeting
History of Mathematics
tredition

- Band 23 (2024): Wolfgang Lange:
Edition des Briefwechsels von Carl Friedrich Gauß (1777–1855) und Johann Friedrich Benzenberg (1777–1846). Hg. von G. Wolfschmidt.
- Band 24 (2014): Wolfschmidt, Gudrun (Hg.):
Kometen, Sterne, Galaxien – Astronomie in der Hamburger Sternwarte. Zum 100jährigen Jubiläum der Hamburger Sternwarte in Bergedorf.
- Band 25 (2016): Wolfschmidt, Gudrun (Hg.):
Wissen aus 400 Jahren Chemie in Hamburg. Hamburgs Geschichte einmal anders, Teil 4.
- Band 26 (2026): Eike-Christian Harden: *Concordia Res Parvae Crescunt – Fortschritte in Naturwissenschaft und Technik im Goldenen Zeitalter der Niederlande.* Hg. von Gudrun Wolfschmidt.
- Band 27 (2014): Susanne M. Hoffmann: *lingua sine limitibus – Analysen zur Sprache der Bilder und Bildsprachen, insbesondere zur Kommunikation von Fachinformationen.* Sprachen der Populärdidaktik mit zwei- bis vierdimensionalen Medien an Beispielen der Astronomie. Hg. G. Wolfschmidt.
- Band 28 (2014): Wolfschmidt, Gudrun (Hg.): *Der Himmel über Tübingen. Barocksternwarten – Landesvermessung – Astrophysik.* Proceedings der Tagung des AKAG in Tübingen 2013.
- Band 29 (2013): Wolfschmidt, Gudrun (Hg.):
Sonne, Mond und Sterne – Meilensteine der Astronomiegeschichte. Zum 100jährigen Jubiläum der Hamburger Sternwarte in Bergedorf.
- Band 30 (2024): Hans G. Beck: *Astrobecks Sternzeiten. Aus dem Leben des Industrie-Astronomen Hans G. Beck.* Bearbeitet und herausgegeben von Gudrun Wolfschmidt.
- Band 31 (2015): Wolfschmidt, Gudrun (Hg.):
Astronomie in Franken. Von den Anfängen bis zur modernen Astrophysik – 125 Jahre Dr. Karl Remeis-Sternwarte Bamberg (1889). Proceedings der Tagung des Arbeitskreises Astronomiegeschichte in der AG in Bamberg 2014.
- Band 32 (2018): Wolfschmidt, Gudrun (Hg.):
Astronomie und Astrologie im Kontext von Religionen. Proceedings der Tagung des Arbeitskreises Astronomiegeschichte in der AG in Göttingen 2017.
- Band 33 (2016): Wolfschmidt, Gudrun (ed.):
Enhancing University Heritage-Based Research. Proceedings of the XV Universeum Network Meeting, Hamburg, 2014.
- Band 34 (2026): Ewering, Christoph: *Bürgerliches Sammeln im 19. Jahrhundert. Das Museum Godeffroy.* Hg. von G. Wolfschmidt.
- Band 35 (2018): Wolfschmidt, Gudrun (Hg.):
Baudenkmäler des Himmels – Astronomie in gebautem Raum und gestalteter Landschaft. Proceedings der Tagungen der Gesellschaft für Archäoastronomie, 2014 bis 2016.

Nuncius Hamburgensis –
Beiträge zur Geschichte der Naturwissenschaften, Band 24
Gudrun Wolfschmidt (Hg.)
Kometen, Sterne, Galaxien
Astronomie in der Hamburger Sternwarte
tredition

Nuncius Hamburgensis –
Beiträge zur Geschichte der Naturwissenschaften, Band 23
Gudrun Wolfschmidt (Hg.)
Sonne, Mond und Sterne
Meilensteine der Astronomiegeschichte
Zum 100jährigen Jubiläum
der Hamburger Sternwarte in Bergedorf
tredition

Nuncius Hamburgensis –
Beiträge zur Geschichte der Naturwissenschaften, Band 42
Gudrun Wolfschmidt (Hg.)
Orientierung, Navigation
und Zeitbestimmung
Wie der Himmel den Lebensraum
des Menschen prägt
tredition

Nuncius Hamburgensis –
Beiträge zur Geschichte der Naturwissenschaften, Band 38
Gudrun Wolfschmidt (Hg.)
Astronomy in the Baltic
Astronomie im Ostseeraum
Proceedings der Tagung des Arbeitskreises
Astronomiegeschichte in Kiel 2015
tredition

Nuncius Hamburgensis –
Beiträge zur Geschichte der Naturwissenschaften, Band 44
Panagiotis Kitmeridis
Popularisierung der
Naturwissenschaften
am Beispiel des
Physikalischen Vereins Frankfurt
tredition

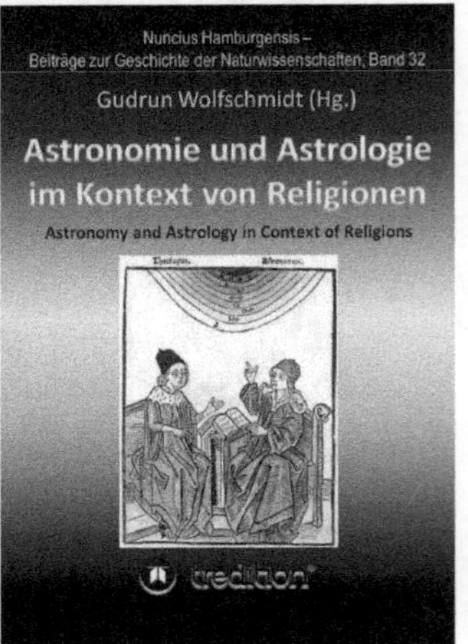
Nuncius Hamburgensis –
Beiträge zur Geschichte der Naturwissenschaften, Band 32
Gudrun Wolfschmidt (Hg.)
Astronomie und Astrologie
im Kontext von Religionen
Astronomy and Astrology in Context of Religions
tredition

Nuncius Hamburgensis –
Beiträge zur Geschichte der Naturwissenschaften, Band 41
Gudrun Wolfschmidt (Hg.)
Popularisierung
der Astronomie
tredition

Nuncius Hamburgensis –
Beiträge zur Geschichte der Naturwissenschaften, Band 35
Gudrun Wolfschmidt (Hg.)
Baudenkmäler des Himmels –
Proceedings der Tagungen der
Gesellschaft für Archäoastronomie
tredition

Irena Kampa
Die astronomischen
Instrumente von
Johannes Hevelius
tredition

- Band 36 (2017): Wolfschmidt, Gudrun (ed.):
Festschrift – Proceedings of the Christoph J. Scriba Memorial Meeting – History of Mathematics.
- Band 37 (2020): Wolfschmidt, Gudrun (Hg.):
70 Jahre Observatorium Hoher List. Sieben Jahrzehnte astronomische Beobachtung in der Eifel.
- Band 38 (2018): Wolfschmidt, Gudrun (Hg.):
Astronomie im Ostseeraum – Astronomy in the Baltic. Proceedings der Tagung des Arbeitskreises Astronomiegeschichte in der Astronomischen Gesellschaft in der AG in Kiel 2015.
- Band 39 (2015): Pfitzner, Elvira:
Vom Jakobsstab zur Spektralanalyse – Astronomie an der Rostocker Universität. Bearb. und hg. von Gudrun Wolfschmidt.
- Band 40 (2015): Kost, Jürgen:
Wissenschaftlicher Instrumentenbau der Firma Merz in München (1838–1932). Bearb. und hg. von Gudrun Wolfschmidt.
- Band 41 (2017): Wolfschmidt, Gudrun (Hg.):
Popularisierung der Astronomie. Proceedings der Tagung des Arbeitskreises Astronomiegeschichte in der AG in Bochum 2016.
- Band 42 (2019): Wolfschmidt, Gudrun (Hg.):
Orientierung, Navigation und Zeitbestimmung – Wie der Himmel den Lebensraum des Menschen prägt. Tagung der Gesellschaft für Archäoastronomie in Hamburg 2017.
- Band 43 (2017): Carlotta Martini: *Zwei Frauenleben für die Wissenschaft im 18. Jahrhundert. Eine vergleichende Fallstudie zu Émilie du Châtelet und Maria Gaetana Agnesi.* Bearbeitet und herausgegeben von Gudrun Wolfschmidt.
- Band 44 (2018): Panagiotis Kitmeridis: *Popularisierung der Naturwissenschaften am Beispiel des Physikalischen Vereins Frankfurt.* Überarbeitet und herausgegeben von Gudrun Wolfschmidt.
- Band 45 (2026): Wolfschmidt, G. (Hg.): *Hamburgs Geschichte einmal anders – Entwicklung der Naturwissenschaften, Medizin und Technik, Teil 5.*
- Band 46 (2024): Wolfschmidt, Gudrun (Hg.):
Vom Abakus zum Computer – Geschichte der Rechentechnik, Teil 2. Mit einem Katalog der Instrumente und Modelle in der Sammlung des GNT.
- Band 47 (2018): Irena Kampa: *Die astronomischen Instrumente von Johannes Hevelius.* Hg. von Gudrun Wolfschmidt.
- Band 48 (2020): Wolfschmidt, G. (Hg.): *Maß und Mythos, Zahl und Zauber: Die Vermessung von Himmel und Erde.* Tagung der Gesellschaft für Archäoastronomie in Dortmund 2018.

Nuncius Hamburgensis –
Beiträge zur Geschichte der Naturwissenschaften, Band 37
Gudrun Wolfschmidt (Hg.)
70 Jahre Observatorium
Hoher List
Sieben Jahrzehnte astronomische
Beobachtung in der Eifel
tredition

Nuncius Hamburgensis –
Beiträge zur Geschichte der Naturwissenschaften, Band 54
Gudrun Wolfschmidt (Hg.)
Erich Meyer:
Auf den Spuren Johannes Keplers
Zum 450. Geburtstag Johannes Keplers
tredition

Gudrun Wolfschmidt (Hg.)
Vom Abakus
zum Computer, Teil 1
tredition

Nuncius Hamburgensis –
Beiträge zur Geschichte der Naturwissenschaften, Band 48
Gudrun Wolfschmidt (Hg.)
Maß und Mythos,
Zahl und Zauber
Vermessung
von Himmel
und Erde
tredition

Nuncius Hamburgensis –
Beiträge zur Geschichte der Naturwissenschaften, Band 55
Gudrun Wolfschmidt &
Susanne M. Hoffmann (ed.)
Applied and Computational
Historical Astronomy
Angewandte und computergestützte historische Astronomie
Proceedings of the Splinter Meeting
in the Astronomische Gesellschaft, Sept. 25, 2020
tredition

Nuncius Hamburgensis –
Beiträge zur Geschichte der Naturwissenschaften, Band 49
Gudrun Wolfschmidt (Hg.)
Internationalität
in der astronomischen Forschung
des 18. bis 20. Jahrhunderts
Internationality
in the Astronomical Research
of the 18th to 20th Century
tredition

Nuncius Hamburgensis –
Beiträge zur Geschichte der Naturwissenschaften, Band 51
Gudrun Wolfschmidt (Hg.)
Himmelswelten und
Kosmovisionen
Imaginationen, Modelle,
Weltanschauungen
Proceedings der Jahrestagung der
Gesellschaft für Archäoastronomie, Gilching 2019
tredition

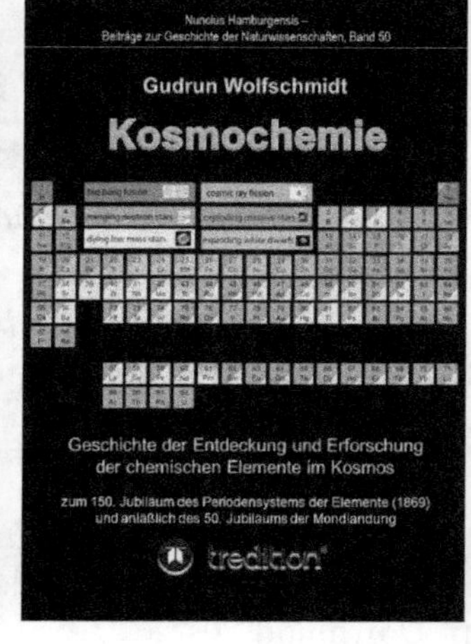
Nuncius Hamburgensis –
Beiträge zur Geschichte der Naturwissenschaften, Band 50
Gudrun Wolfschmidt
Kosmochemie
Geschichte der Entdeckung und Erforschung
der chemischen Elemente im Kosmos
zum 150. Jubiläum des Periodensystems der Elemente (1869)
und anläßlich des 50. Jubiläums der Mondlandung
tredition

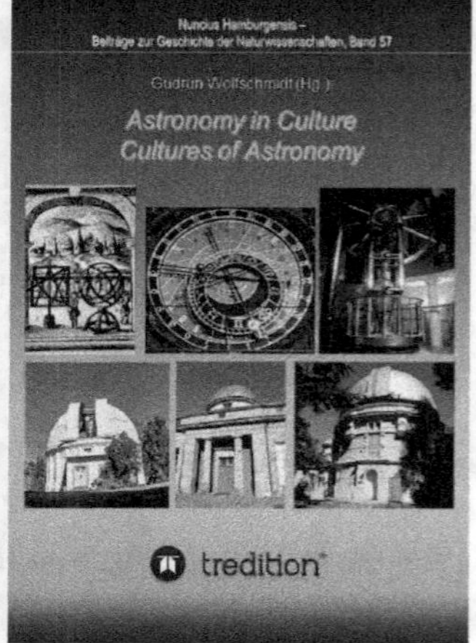
Nuncius Hamburgensis –
Beiträge zur Geschichte der Naturwissenschaften, Band 57
Gudrun Wolfschmidt (Hg.)
Astronomy in Culture
Cultures of Astronomy
tredition

- Band 49 (2020): Wolfschmidt, Gudrun (Hg.): *Internationalität in der astronomischen Forschung (18. bis 21. Jahrhundert).* Proceedings der Tagung des Arbeitskreises Astronomiegeschichte in Wien 2018.
- Band 50 (2022): Wolfschmidt, Gudrun (Hg.): *Kosmochemie – Geschichte der Entdeckung und Erforschung der chemischen Elemente im Kosmos.* Zum 150. Jubiläum des Periodensystems der Elemente (PSE, 1869). und anläßlich des 50. Jubiläums der Mondlandung. Proceedings der Tagung des Arbeitskreises Astronomiegeschichte in der AG in Stuttgart 2019.
- Band 51 (2020): Wolfschmidt, G. (Hg.): *Himmelswelten und Kosmovisionen – Imaginationen, Modelle, Weltanschauungen.* Proceedings der Tagung der Gesellschaft für Archäoastronomie in Gilching 2019.
- Band 52 (2023): Wolfschmidt, Gudrun (Hg.): *Instrumente, Methoden und Entdeckungen für innovative Entwicklungen in der Astronomie.* Proceedings der Tagung des Arbeitskreises Astronomiegeschichte in der AG in Bremen 2022.
- Band 53 (2024): Wolfschmidt, Gudrun (Hg.): *Kosmische Geometrien und Symmetrien – Welterfahrung, Architektur und Landschaft.* Proceedings der Tagung der Gesellschaft für Archäoastronomie in Weimar 2023.
- Band 54 (2021): Erich Meyer: *Auf den Spuren Johannes Keplers – Zu seinem 450. Geburtstag.* Bearbeitet und hg. von Gudrun Wolfschmidt.
- Band 55 (2021): Wolfschmidt, Gudrun & Susanne M. Hoffmann (ed.): *Applied and Computational Historical Astronomy.* Proceedings of the Splinter Meeting in the AG, Sept. 25, 2020.
- Band 56 (2021): Christoph Prignitz: *Zeit für Hamburg – Eine Uhr der Sternwarte und ihr historisches Umfeld.* Mit Beiträgen und herausgegeben von Gudrun Wolfschmidt.
- Band 57 (2022): Susanne M. Hoffmann & Gudrun Wolfschmidt (ed.): *Astronomy in Culture – Cultures of Astronomy.* Proceedings of the Splinter Meeting in the AG, Sept. 14–16, 2021.
- Band 58 (2023): Wolfschmidt, Gudrun (Hg.): *Wandlungen in Raum und Zeit: Himmel – Heimat – Weltverständnis.* Proceedings der Tagung der Gesellschaft für Archäoastronomie in Recklinghausen 2022.
- Band 59 (2024): Wolfschmidt, Gudrun (Hg.): *Astrophysik seit 1900 – Jubiläum von Karl Schwarzschild (1873–1916) und Ejnar Hertzsprung (1873–1967).* Proceedings des AKAG in Berlin 2023.

- Ziegler, Wolfram: *Aus dem Leben eines Franken. Dr. August Ziegler (1885–1937) – Pflanzenzüchter in Togo und Rebenzüchter in Bayern.* Hg. von Gudrun Wolfschmidt. 2017.
- *Harmony and Symmetry. Celestial regularities shaping human culture.* Ed. by SONJA DRAXLER, MAX E. LIPPITSCH & GUDRUN WOLFSCHMIDT. Hamburg: tredition (SEAC Publications, Vol. 01) 2020.

Abbildung 12.1:
Berliner Wappen

(Foto: Gudrun Wolfschmidt)

Personenregister

www.ingramcontent.com/pod-product-compliance
Lightning Source LLC
LaVergne TN
LVHW080556200726
843510LV00004B/934

* 9 7 8 3 3 8 4 4 4 6 3 4 3 *